SEA & AIR

The Marine Environment

Second Edition

SEA & AIR

The Marine Environment

Second Edition

JEROME WILLIAMS
Research Professor of
Environmental Protection
U.S. Naval Academy

JOHN J. HIGGINSON
Commander, U.S. Navy

JOHN D. ROHRBOUGH
Commander, U.S. Navy

NAVAL INSTITUTE PRESS
Annapolis, Maryland

Second printing, 1975

Cover photograph: Robert deGast

Unless otherwise indicated by a
credit line, all photographs are
official U.S. Navy.

Library of Congress Catalogue Card Number: 72-93196
ISBN: 0-87021-596-5

Printed in the United States of America

To Lelia, Nancy, and Sylvia

Critics par excellence

Preface

Throughout its investigations the Panel has been impressed by the unity of environmental sciences.

The President's
Science Advisory Committee
June, 1966

Findings similar to those of the President's Science Advisory Committee prompted the writing of this book. The intimate interaction of meteorology and oceanography, the similarity of their physical laws, their parallel structures—all directed that we approach the environmental sciences from a single viewpoint. Wherever it appeared that the sea and the air could best be studied together, we have so studied them. The result is a departure from normal discussions of meteorology and oceanography.

Because the subject is interdisciplinary, no attempt has been made to standardize the units of measure used. Since diversity of measurement has always been a trademark of the field, the ability to make mental cross-conversions is obviously an asset for any student of the environmental sciences. Therefore, we have used the units most commonly used for measuring each parameter. By keeping the mathematics requirements to a minimum, we have attempted to produce material comprehensible to the average college student, while still retaining some measure of intellectual challenge.

The first edition of *Sea and Air*, subtitled *The Naval Environment*, was designed primarily for use in an introductory course given to all midshipmen at the U.S. Naval Academy. However, the needs of anyone interested in the marine environment are not much different from those of the Navy and for that reason, as well as to reflect its broader base, this second edition is subtitled *The Marine Environment*. The first chapter has been re-written and a new last chapter, summarizing pollution problems in the sea and in the air, has replaced the one that dealt with the reasons why it was incumbent on naval officers to know their environment.

It is our hope that, from this edition even more than from the first, the reader will acquire an awareness of the myriad ways the environment affects day-to-day human behavior. Some activities are dependent upon the weather. Others are greatly modified by the state of the sea. Almost all, to some degree, are related to the behavior of the fluid medium. Today's technology has given man the ability to probe the vastness of the skies and the mysterious depths of the seas. We hope to impart a basic understanding of sea and air, and of the impact of their interaction in the marine environment.

J. W. J. J. H. J. D.R.

U.S. Naval Academy
Annapolis, Maryland

Foreword to the First Edition

Publication of this book is unique in itself, inasmuch as it combines basic information about both the air and sea along with the physical parameters and processes which cause each to act and react in certain ways. Until this time a great deal of such related information has been hidden away within the minds and notes of the relatively small number of persons who had studied both sides of the sea-air interface.

It is obvious that the authors should call upon their previous background and experience, in the academic world as well as in military operations, as aids in preparing the copy. The work of Professor Williams has been known within the oceanographic community for some time. The military authors add a rather nice combination of experience, inasmuch as Lieutenant Commander Rohrbough is a surface officer and Lieutenant Commander Higginson a qualified naval aviator. Both are graduates of the Navy's initial program to teach officers about the air-ocean environment at the Naval Postgraduate School in Monterey, California. Their eventual assignment to the Naval Academy as instructors led to the concept of this book which has now been prepared in concert with their academic associate, Professor Williams. The initial attempt to produce this volume was for the express purpose of compiling information, which could not be found in any one publication or text, for use by midshipmen at the U.S. Naval Academy at Annapolis. The success of these authors in combining the many facets of the ever changing air-water world into a single text will be evident to anyone who scans the bibliographies and reading lists appended to each chapter.

I would hope that this latest publication from the Naval Institute might result in an increase in both basic knowledge and a fuller appreciation of the subject by those students, laymen, and yes, even Navy men, who continue to show an ever increasing interest in our environment.

R. W. HAUPT
Captain, U. S. Navy
Staff, Oceanographer of the Navy

Table of Contents

Symbols

Symbol	*Name*	*Quantity Represented*
a		absorptivity
A		albedo, wave amplitude
A		arctic air
c		wave speed
c		continental air
cgs		centimeter-gram-second
cm		centimeter
C		coriolis force
°C	degree Celsius	temperature
Cl°/₀₀		chlorinity in parts per thousand
d		water depth
db	decibel	sound level

Symbol	*Name*	*Quantity Represented*
D		depth of frictional influence, duration of wind
e		vapor pressure, emissivity
e_s		saturation vapor pressure
E		energy emitted per unit area per unit time, total wave energy of sea surface
°E	degrees East	longitude
E		equatorial air
f		wave frequency
F		frictional force, fetch
°F	degrees Fahrenheit	temperature
g		acceleration of gravity
gm	grams	mass or weight
h		fluid column height
H		water wave height
$\bar{H}$		average wave height
$H_{1/3}$		significant wave height
Hz	hertz	frequency
I		intensity
km	kilometer	distance
kts	knots	speed
k		cold air
°K	degrees Kelvin	temperature
l		length
L		wavelength
mbs	millibars	pressure
mm	millimeters	length
m		maritime air
°N	degrees North	latitude
p		pressure
P		pressure gradient force
P		polar air
r		radius of curvature
R.H.		relative humidity
S		centrifugal force
°S	degrees South	latitude
S‰		salinity in parts per thousand
t		time
T		gravitational attractive force, temperature, wave period, absolute temperature in °K
T		tropical air
v		linear speed

Symbol	*Name*	*Quantity Represented*
w		warm air
W		wind speed, wind stress force
°W	degrees West	longitude
α	alpha	absorption coefficient
γ_d	gamma *d*	dry adiabatic lapse rate
γ_m	gamma *m*	moist adiabatic lapse rate
ε	epsilon	wave phase
η	eta	displacement from equilibrium position
λ	lambda	electromagnetic energy wavelength
£	libra	characteristic absorption length
$£_a$	libra *a*	characteristic attenuation length
$£_s$	libra *s*	characteristic scattering length
ρ	rho	density
$\bar{\rho}$	rho bar	average density
σ	sigma	Stefan-Boltzmann constant, density anomaly
σ_t	sigma *t*	density anomaly when pressure is not a factor
ϕ	phi	latitude
Ω	omega	earth's angular rotational speed

SEA & AIR

The Marine Environment

Second Edition

CHAPTER ONE

Prologue

101 On the Sea. In mid-April two merchant tankers bound from Bremerhaven to Philadelphia were preparing to sail from the west opening of the English Channel. The northern route, which follows closely the great circle and is, therefore, the shortest, was recommended by a ship routing service in Norfolk, Virginia, in spite of the fact that in April violent weather is normal along that route. The master of one of the tankers accepted this recommendation, while the master of the other, having little faith in "new-fangled shore-based navigators," decided to stick to the traditional route and headed southwest.

The ship that used the prescribed routing had a very pleasant crossing. The winds never exceeded 15 knots and the seas were following most of the time. As a result of the fine weather, she sailed into port on schedule, having been at sea only nine days.

Seven hundred miles out in the Atlantic, the second ship encountered 35-knot head winds and heavy seas. In order to evade a storm he thought was to the north, her master directed her farther to the south, by which time she was making turns for only 6 knots, and the storm's ferocity was increasing. On the day the first tanker reached port, the second one was at mid-ocean and conditions were not improving. She finally reached port a full ten days later. That extra time at sea was costly in fuel, salaries, and damage repair. Her master then realized that he should take advantage of the new weather navigation system.

Wave-forecasting, which makes it possible to direct ships through the calmest waters, was developed and pioneered by the U.S. Navy in the 1940s, refined in the 1950s, and is presently used to direct all ships under the control of the Navy. There are also many commercial ship-routing services, and private shipping companies apparently consider the cost of such services is more than offset by savings of time and by minimal damage to cargoes and vessels.

Although the real incident described above, with its ten-day time loss, is an extreme case, there have been many similar ones in the last two decades. It has been estimated, for example, that for all ships and at all times of the year the average saving realized by ship routing on a North Atlantic passage is seven hours. It is obvious that a knowledge of the

Courtesy: B. J. Nixon

environment and the ability to use that knowledge are great benefits to the mariner, whether he be aboard a merchant vessel, a military ship, or a pleasure craft.

102 In the Clouds. The pilot of a high-flying jet aircraft is well aware of the influence that the environment has upon his flight. As a frequent traveler in the upper regions of the atmosphere, he must be aware of many hazards. At 45,000 feet, without a supplementary supply of oxygen, a person's life is measured in seconds.

A pilot has to be keenly aware of the winds that may assist or reduce his speed over the ground. The jet stream, with winds in excess of 150 knots, is a force to be reckoned with. For maximum performance consistent with the direction of flight, this high-altitude wind must be carefully avoided or judiciously utilized. Commercial airlines flying across country fly at one altitude when traveling east and at another when traveling west: in this manner, the jet stream adds to the aircraft's speed in one direction and does not detract from it in the other.

Before they complete flight training, most aviators learn that the rate of heat acceptance by the earth's surface is far from uniform. They also learn that the convective lift from warm earth does not exist over the ocean and that, consequently, when a plane must pass over a stretch of ocean as it comes in for a landing it will, while over the water, require a small increment of power; if, however, the increment is not removed when the plane passes from over-water to over-land, the approach will be high at the critical point of landing.

Thunderstorm clouds, although beautiful in appearance and majestic in structure, are feared by the aviator. They are caused by and contain vertical motions that can throw an aircraft out of control. Quite often, icing conditions and hail are encountered within these clouds, and the associated lightning has been known to damage outer parts of aircraft. Even those responsible for the safety of aircraft on the ground must be alert for severe storms that could cause damage to planes not properly secured. At times, aircraft may have to be flown to an area out of the reaches of the storm.

Pilots may also encounter clear-air turbulence (CAT)—turbulence whose existence is not suggested by the presence of clouds. Such turbulence can cause serious structural damage to an aircraft and can even cause it to crash. Clear-air turbulence is under study, but at this time only limited success in predicting its presence has been achieved.

Another thing pilots learn is that a runway in close proximity to the sea can constitute a severe hazard. Currents from a certain direction cause water from cold depths to be brought to the surface. This upwelling of cold water causes severe fogs, which make a landing difficult or impossible.

At the point of landing, a pilot is conscious of the fact that he is attempting to maneuver his plane and make a transition between two media which are in motion in relation to one another. A cross-wind landing taxes the skill of any pilot. The navy carrier pilot faces probably the most challenging of all aviation situations. He must translate motion in a moving air mass to a ship moving on the ever-changing surface of the sea. He must know the environment, for one mistake can cause loss of lives and extensive property damage.

Radar is a key factor in successful flying, but the degree to which it is useful is closely tied to atmospheric conditions. As often as not, radar is used for the purpose of making a safe landing and, since precipitation tends to obscure portions of the radarscope and thus to mask the approach to a field, care must be used in bad weather. As a sound wave is bent in the liquid medium, so is radar bent in the gaseous medium. To use this tool properly, it is necessary to be aware of effects created by the environment and to know how to compensate for them.

103 Below the Waves. A vehicle that moves in a single-fluid environment can be operated much more efficiently than one that operates at the interface between two fluids. Thus, submarines and airplanes, both of which do operate within single-fluid environments, use their propulsion energy efficiently. The submarine, for example, does not have to concern itself with the extreme motions experienced, especially during storm periods, at the air-sea interface, because, generally speaking, motional activity decreases with depth. The submarine has the further great advantage of operating within a weather environment that is essentially constant. Oceanic temperature varies somewhat, but not to anything like the extent surface temperature does.

This last advantage has particular significance with regard to the desired transportation of oil from the frigid Arctic waters off the north coast of Alaska to where it can be used. It has been suggested that large tanker submarines, or submarines pulling trains of submerged tank barges, might be used for this purpose. This may be a much more practical method of transportation than using conventional surface tankers. The idea is especially valid since the converted supertanker *Manhattan* had so much trouble negotiating the northwest passage even during the spring period, when the sea ice was not at maximum coverage.

Of course, there are some drawbacks to the commercial use of submarines for moving cargo. For example, the deep ocean is not as placid as it might seem. Deep-ocean turbulence (DOT), a situation similar to clear-air turbulence, is sometimes found in the ocean, and it is extremely difficult to forecast the occurrence of this potentially dangerous phenomenon.

Internal waves, sometimes more than a hundred feet high, create another hazard. Even though such waves have periods on the order of magnitude of 15 to 20 minutes, they present as much difficulty for a submarine as surface waves of similar height do for a surface ship. Nevertheless, large numbers of mammoth submarines might very well be used in the future for transporting cargo. If this is to be the case, some progress will have to be made in solving the problems of forecasting DOT and internal waves.

Bathymetric navigation, or the use of bathymetric features for determining position, requires more knowledge of the topography of the ocean bottom than is presently available. However, when this science is further developed, undersea navigation will be considerably aided.

104 Back on Land. The air-sea environment is of interest to all people, whether or not they are actively engaged in working in or with it. The weather, for example, certainly affects everyone, from the farmer whose crops depend upon it, to the sports fan whose plans are often changed by it. The ability to predict with confidence for more than about eight hours in advance is something that meteorologists have been striving for continuously.

It is well known that knowledge of the physical processes occurring at the air-ocean interface is of prime importance in making long-term weather predictions. The body of this knowledge is increasing very rapidly, but it still has not reached the point where accurate long-range weather predictions can be made.

The environment is also of interest in terms of what we use it for, one such thing being waste disposal. For example, everything we burn, from charcoal for roasting hot dogs to gasoline for powering our automobiles, discharges waste products into the atmosphere. Many solid waste products, domestic sewage, industrial wastes, and excess heat from electric power plants, are discharged into the water. Waste-disposal problems are becoming acute, as the volume of waste products increases each year, while the volume of atmosphere and ocean into which they are discharged remains the same.

Another use we make of the environment is for our recreational activities, such as swimming, fishing, and boating, and these are directly affected by our waste-disposal practices.

In order intelligently to manage our environmental resources, we must know something about them. The following chapters attempt to provide the reader with sufficient knowledge to enable him to understand some of the management decisions that have been made and to protect himself from the vagaries of his ever-changing environment.

CHAPTER TWO

Earth: The Foundation

201 Origin of the Mass. The earth is estimated to be between 4.5 and 6 billion years old. When formed, primitive earth was unlike today's land, sea, and air. This chapter will provide an insight into the terrestrial origins and then will lay a geologic foundation for the primary study, the interactions between the two vital fluids—the sea and the air.

Several theories which describe the earth's formation had been postulated and then progressively replaced as new evidence became available. The first rational theory was presented by Pierre Simon Laplace, the French astronomer and mathematician, in 1796. The *nebular hypothesis*, as it was called, postulated a large gaseous nebula in rotation. The gaseous mass cooled and shrank; this shrinkage caused the rotational velocity to increase. As the increased velocity conserved the rotational momentum, hot globs began spinning off in successive rings. They cooled to form the present planets moving around the remaining mass, or, the sun. This theory, sketchily presented, has long been held in disrepute

Waipio Gulch, Hawaii

but does stand as a very imaginative first attempt based on the state of physics and astronomy at that time.

Another theory, known as the *hypothesis of tidal disruption*, was postulated by T. C. Chamberlin and F. R. Moulton in the beginning of the twentieth century.

Essentially the sun was thought to have almost collided with another large star as the two passed through space. The tidal gravitational forces, caused by the changing proximity of the two molten bodies, pulled a succession of hot masses from the sun. These assumed nearly circular orbits, cooled and became the planets. This theory has also fallen into disfavor for various reasons.

The most promising theory now enjoying wide acceptance is called the *dust cloud hypothesis*. Since 1944 C. F. von Weizsäcker and G. P. Kuiper have developed the basic theory on the belief that the solar system evolved from dust clouds. Such dust clouds are observed constantly throughout the galaxy with modern telescopes. Accelerations and decelerations of the dust particles occurred at different radii around a cold nebulous center. These angular velocity changes allowed collisions and caused turbulent eddies to form pockets of increased mass density.

The protoplanets, or pockets of mass accumulation, were diffuse bodies much larger than the present planets. The solar system, as known today, came into being as the planets contracted and solidified. The sun was formed when the central mass shrank and became heated because of the increased internal pressure of the contracting matter. The identical process is believed to be responsible partially for the extremely high temperature of the earth's interior. The dust cloud theory has shortcomings but many of its facets have been held quite acceptable in their present form.

202 Origin of the Fluids. Atmospheric and hydrospheric developments were undoubtedly complex. The great gaseous fluid, enveloping the protoplanet, is believed to have consisted of hydrogen, helium, ammonia, and methane. Scientists have deduced that much of the earth's original hydrogen and helium was lost to space and that some of the present constituents—oxygen, more hydrogen, nitrogen, and carbon—were conserved in relatively non-volatile compounds to be spewed out from the interior.

When the stable compounds were brought to the surface via volcanoes and hot springs, the last formation stages began. Original di-hydrogen oxide gas (water vapor) went directly to the atmosphere to be disassociated by solar radiation into oxygen and hydrogen. The latter was lost to space. The new free oxygen reacted with residual and newly released ammonia to produce water and free nitrogen ($4NH_3 + 3O_2 = 4N + 6H_2O$).

Residual methane with oxygen formed carbon dioxide and water ($CH_4 + 2O_2 = CO_2 + 2H_2O$). Other more complex chemical reactions also occurred to give the atmosphere its present constitution.

Continued volcanic action probably brought the water and some of the dissolved salts for the hydrosphere from the earth's interior. This extrusion continued for millions of years and is still occurring today. The oceans themselves filled as the atmosphere's ambient temperature dropped below 100°C (Celsius).

In brief, changes of great magnitude were required to produce today's livable fluids. The planets had reached their relatively stable states by the beginning of the Cambrian Age (six hundred million years ago).

203 The Earth's Interior Structure. In the study of the interior structure one must understand how solid and liquid materials transmit seismic (produced by earthquakes) waves. When the vibrations of a rock affect all surrounding rocks, particular waves are transmitted concentrically in two types, a compressional wave and a shear wave. The compressional wave (primary or P wave) travels equally well through solids and liquids, but at a faster rate than the shear. The shear wave (secondary or S wave) travels well through solids but is not easily detectable at great distances through liquids, since liquids are incapable of supporting shear.

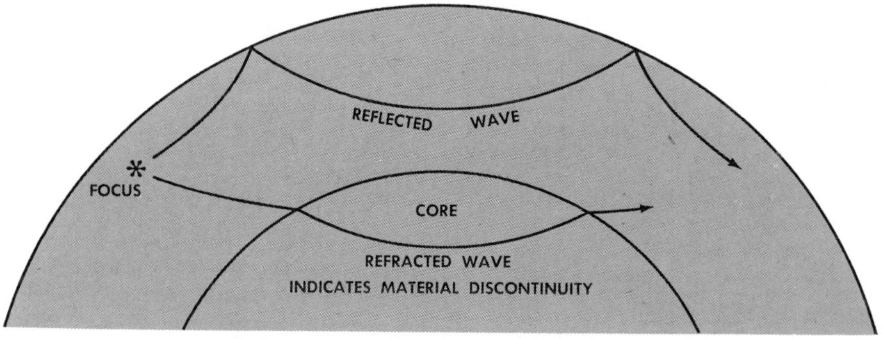

Figure 2-1 Seismic-Wave Transmission

By studying the seismic wave records, scientists have discovered and verified that various and distinctly homogeneous layers exist below the outer crust. The earth probably has a core of nickel-iron, the inner section of which is believed to be solid due to the extreme interior pressure. The solid inner section has been detected by the refractions of the P waves as they penetrate the general core. The temperatures of the inner and outer cores are believed to be the same; the temperature has been held constant by radioactivity in the outer layers whose energy release balances the small heat flow through the ocean bottom (Figure 2-2).

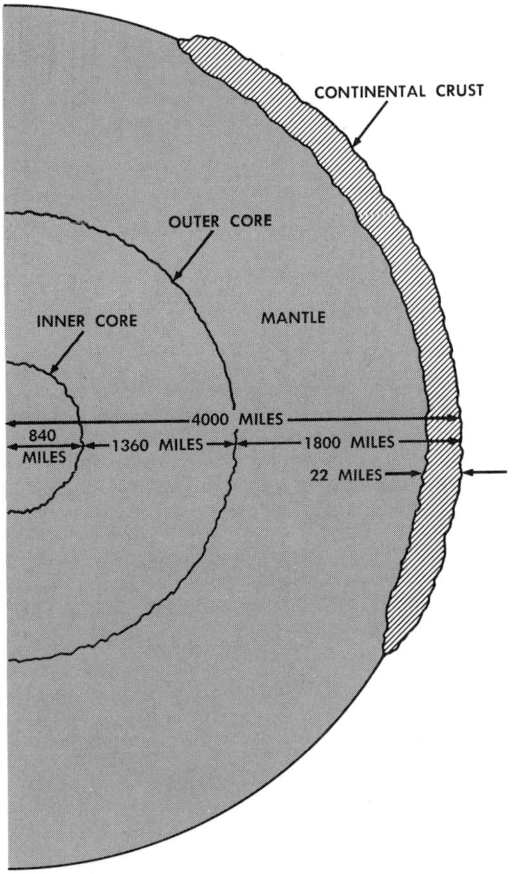

Figure 2-2 The Earth's Interior

The layer outside the liquid core is called the *mantle*. It is composed of a partially molten dense silicate rock material. The mantle with its great thickness comprises over one-half the earth's total volume.

Above the mantle is the continental and oceanic crust. The crust is separated from the mantle by a thin zone known as the *Mohorovicic discontinuity* after its discoverer, a Yugoslav seismologist, A. Mohorovicic.

The *Moho*, as it is called, is unique because it has an average depth of 35 kilometers (km) under the continents but only a 6-km depth beneath the ocean basins.

Although the *Mohole* (hole to the Moho) *Project* was terminated in 1966, a substitute program to sample the oceanic crust has been active. *The Ocean Sediment Coring Program* started field operations in August 1968, and since that time cores have been recovered from stations in the Atlantic, Pacific, Mediterranean, Caribbean, and Gulf of Mexico. In the first three years of operation, approximately 37,000 feet of core have been retrieved from a total penetration of 210,000 feet at more than 150 separate locations. The most important evidence so far unearthed indicates strongly that the sea floor is spreading at the rate of about 0.5 inches per year (*see* below).

The continental crust is composed of two basic rock types—*sial*, a silicon-aluminum base material, and *sima*, an iron magnesium composite. Sialic rock is generally granitic, while simatic rock is basaltic.

For further illustration of the basic structural differences between the oceanic and continental sections, refer to Figure 2-3. Note that the *sima* layer is continuously present under the oceans and the continents, while the *sial* stops at the continental boundary. Also observe that the depth of the crusted surface increases as the continental area thickens to reach a maximum under the mountain-range sector.

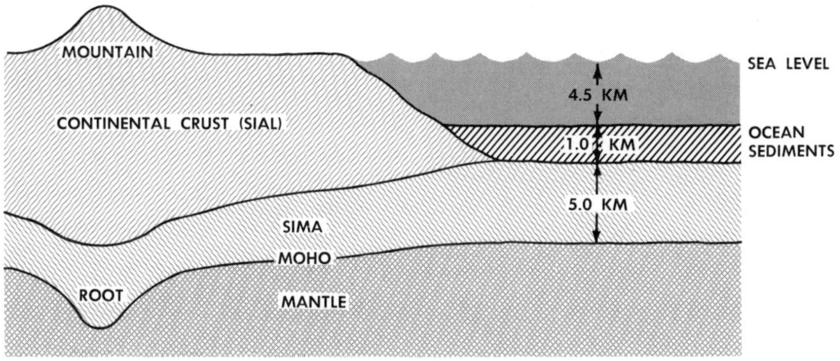

Figure 2-3 Oceanic Crust Versus Continental Crust

Geologists agree that basic differences have existed between the basins and the continents throughout geologic time. Several theories have evolved to explain the formation of the earth's existing natural geography. Some geologists argue the permanency of the oceans' basins, on the premise that they have been more or less in their present positions since early pre-Cambrian days, and they have been subjected to only the small movements caused by local earthquakes and volcanic action.

Others advocate the continental drift idea which theorizes that the land masses were originally welded together in one or two great prehistoric continents. A glance at a globe suggests that eastern South America might fit into the concave section of Africa, for instance. Continuous fittings of the 1,000-fathom curves indicate an amazing fit well north of the Gulf of Mexico and south of the La Plata River. As indicated above, modern evidence lends strong support to this concept.

What causes these masses to be in motion? The *mantle convection current theory* presents the most plausible answer. R. S. Dietz suggests a cellular motion (Figure 2-4), causing movement in the mantle, volcanism at critical sections, and mid-ocean ridges described in the topography section.

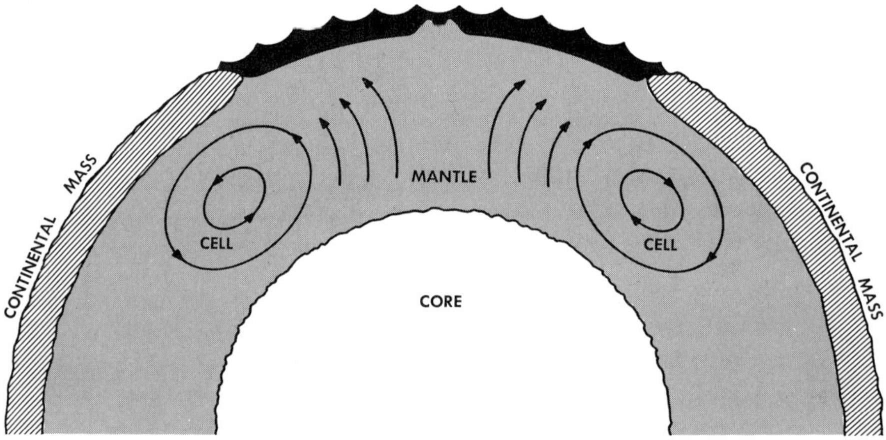

Figure 2-4 Convection Current Theory

The volcanic section or "belts," which incidentally somewhat coincide with the earthquake-epicenter belt, are prevalent around the border of the Pacific Ocean—the Ring of Fire—and along the Alpine-Himalayan belt. The Pacific belt has in excess of 90 historic volcanoes; the Alpine-Himalayan belt has more than 50 extending from the Madeira Islands (eastern North Atlantic) through the Mediterranean and Iran to the Lesser Sunda Islands (north of Australia). The only recent and present volcanic activity in the Atlantic appears in the Lesser Antilles (Caribbean Sea), the Southern Antilles (south of South America) and along the mid-Atlantic Ridges (Azores and Iceland) to the Cape Verde Islands.

204 Isostatic Equilibrium. As depicted in Figure 2-3, the mantle is deeper under the continents—especially the mountains—than under the ocean basins. This stability of the crust-mantle system is explained by the principle of isostatic equilibrium. In very simple form, the principle states that the crustal volume is floating on the semi-plastic mantle, not unlike a block in water. In this analogy, the greater the volume of the block, the more depth the "root" of the block must have. Continental rocks with their greater heights above sea level will have equally greater depths or "roots" than their ocean-basin structure. For instance, the Himalayas are believed to have a root of 70 km to support the great heights (mass) that exist above sea level.

205 Oceanic Dimensions. Until the nineteenth century very little was known about the ocean depths. Mariners had shown little interest in taking soundings in deep water; no equipment existed capable of measuring the depths. During the last century mechanical soundings recorded various depths. The transatlantic cable problem in the 1850's demonstrated to the world that man required knowledge of the nature and depths of the great ocean expanse. The Challenger Expedition in 1872 discovered the Marianas Trench—still noted as the greatest known depth in the ocean.

Using the principles of sonar, technicians have now developed sophisticated depth recorders with accuracies of .05 per cent to 4,000 fathoms and lesser but acceptable accuracies to the greatest depths. The primary problem of correctly mapping the ocean bottom has been that of precise navigation. Now, by the use of navigational satellites, survey ships can ascertain their positions to within 0.2 miles. This great breakthrough should provide many startling discoveries and verifications in the future.

During the third quarter of the nineteenth century, as oceanographers collected soundings from all over the world, cartographers began preparing bathymetric charts. Matthew Fontaine Maury prepared a basic chart of the South Atlantic as early as 1854. Bathymetric charts are drawn with lines called *isobaths* connecting points of equal depth. Isobaths are underwater contour lines which, in concert, pictorially describe the bottom topography. A good navigator is always aware of these isobaths—especially in shallow waters.

Topographic relief depicts all levels on the earth's surface. Mount Everest extends upward 8,850 meters and the Challenger Deep downward 10,915 meters from sea level. This surface relief is only 0.33 per cent of the total 4,000-mile earth radius and the entire crust is only a thin skin covering this gigantic world mass.

A short observation of topographic relief statistics is in order. The frequency distribution of isobaths as shown in Figure 2-5 shows two distinct maxima of relative heights—one at 100 meters and the second at 4,950 meters—using sea level as reference. This nonrandom, bimodal distribution lends strong statistical evidence to the assumption of a

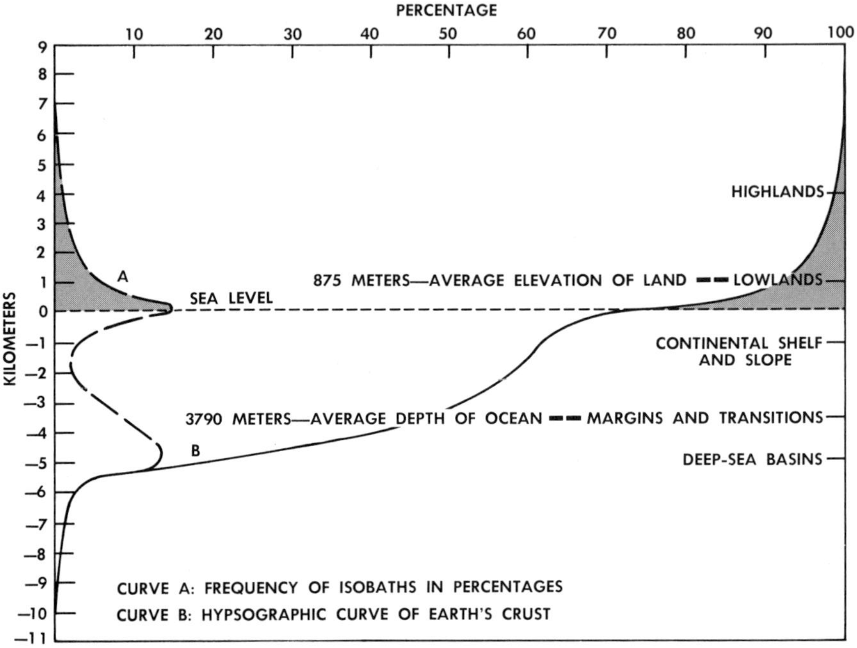

Figure 2-5 Hypsographic Curve of Earth's Crust

basic geologic difference between the oceanic and continental crust. If the depths or heights were distributed completely at random, the result would be a familiar Gaussian, or "bell," curve with a single mode representing the most common level. That level would theoretically be the mid-point between the extreme levels, thereby indicating no basic difference in structure anywhere on earth. In addition, many questions can be resolved utilizing information from the hypsographic curve. For example, the depth capability of a deep submersible may be determined in terms of percentage of ocean-bottom coverage desired. Noteworthy is that of the 70.8 percent of the earth's surface covered by water, only 1 percent has depths exceeding 6,000 meters whereas 5.5 percent has depths less than 200 meters on the continental shelf.

206 Ocean and Sea Boundaries. The oceans are considered to be three in number: the Atlantic which in the south extends from the meridian of Cape Agulhas (20° E) to the line connecting Cape Horn to the Palmer Peninsula and northward to include the North Polar Sea; the Pacific which in the south goes from the Cape Horn line westward to the meridian of the south cape of Tasmania (147° E) and then northward to the Bering Straits; and the Indian Ocean which lies between the other two. This divisional process is made after considering geography, geologic differences, and surface and subsurface current flow, as each applies to the respective oceans.

Mediterraneans are extensions of their respective oceans. The European, American (Caribbean Sea) and Arctic (Arctic Ocean) mediterraneans are part of the Atlantic: the Baltic Sea, Hudson Bay, Persian Gulf, and the Red Sea are considered intracontinental mediterraneans, while the Australian-Asiatic mediterranean is in the Pacific.

The North Sea, Gulf of St. Lawrence, and East China Sea, as examples, are considered *marginal seas* in that they form merely an indentation of their respective oceans into the continental coasts.

207 Bottom Topography. Mariners are aware of the land masses which they see or touch especially when their ships run aground. Until recently those bottom features which do not affect or hazard navigation were of little interest, but now much more is known, catalogued, and studied. The recent international conferences and treaties concerning the countries' ownerships of the various continental shelf areas have served to demonstrate the need for the naval officer to take cognizance of the oceans' basins and their features.

208 Continental Margins. Oceanographers are currently examining closely the continental margins, which consist of the continental shelf and the continental slope (Figure 2-6). The broad area from the intertidal zone to the shelf break is known as the *continental shelf.* It has an average width of 42 miles but varies in width from less than a mile off

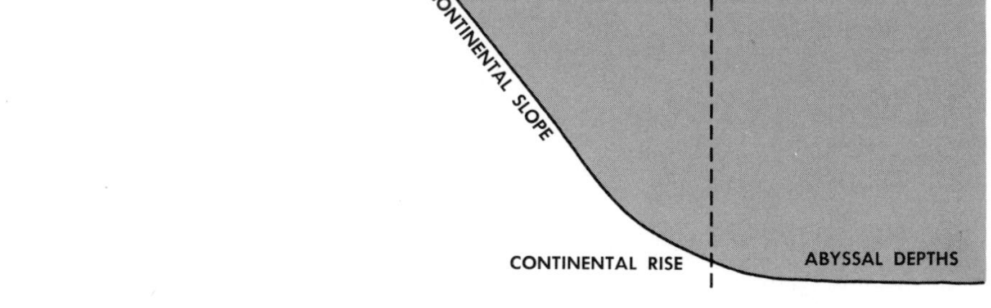

Figure 2-6 Schematic Section of the Continental Margin

the California coast to 800 miles along the north coast of Siberia. The shelves have an average inclination of 10 feet in a mile and have depths at the shelf breaks (seaward boundary) which vary from 50 meters (165 feet) to a maximum of 460 meters (1500 feet). There is presently a great effort to legally define the maximum shelf depth in order to maintain international harmony in the field of bottom exploration.

The *continental slope*, seaward of the *shelf break*, has normal inclinations ranging from 1.5° to 30° with an average near 5°. The lower section of the slope where it gently declines into the abyssal depths is called the *continental rise*

Submarine canyons are permanent features on the shelf and slope. These canyons, such as the Hudson Canyon and Congo Canyon, which actually extend into the estuary for about 15 miles may be associated with river mouths. Other submarine canyons, such as the Tokyo Canyon, Scripps Canyon, and the huge Monterey Canyon, are off bays; while still others have no apparent connection with any continental topographic features.

There are two opposing theories which describe canyon formation. The first theory holds that they are drowned river valleys. The second

Table 2-1 Major Oceanic Trenches

PACIFIC OCEAN	FEET	METERS
Marianas Trench (deepest)	36,204	11,035
Marianas Trench (Challenger Deep)	35,800	10,915
Tonga Trench	35,597	10,853
Kurile-Kamchatka Trench	34,580	10,543
Kermadec Trench	32,809	10,003
Japan Trench	32,153	9,800
North Solomons Trench	29,987	9,142
New Hebrides Trench	29,643	9,038
Yap Trench	28,216	8,602
Palan Trench	26,706	8,142
Peru-Chile Trench	26,427	8,057
Aleutian Trench	25,165	7,672
Nansei Shoto Trench	24,639	7,512
Middle America Trench	21,851	6,662
ATLANTIC OCEAN		
Puerto Rico Trench	30,184	9,392
South Sandwich Trench	27,100	8,262
INDIAN OCEAN		
Sunda Trench	24,442	7,252

theory holds to the premise that over the centuries, turbidity currents scoured and scraped deeper the canyon walls.

Turbidity currents or *density currents* are flows caused by the spontaneous liquefaction of minute particles which then flow down-hill in suspension. As the flow slows, the larger particles filter out successively until only colloids are left. Transoceanic cables have been broken one after another by these turbidity currents which were measured at speeds up to 55 knots. These currents can be caused by earthquakes or by unstable piles of unconsolidated matter, falling down an incline like an avalanche.

Continental borderlands are areas that have irregular bottom features where the shelf break is not easily detected, such as the area near the coast of Southern California around Catalina Island. Another feature worthy of note, having continental crustal properties, is typified by the intermediately deep plateau regions with steep escarpments, such as the Blake Plateau off the coast of Georgia.

The last important feature of the continental margin is the *marginal trench*, which is found in close proximity to the shelf-slope region. Notice that the island arcs in the Pacific, in many cases, have associated trenches tending to follow the curve of the arc. These omnipresent similarities suggest possible related formation factors. The important trenches are listed in Table 2-1.

209 The Ocean Basin. The *ocean basin* is the vast expanse of a complete basaltic structure beneath the consolidated and unconsolidated sediments. The *abyssal plain* is that region at great depths which is smooth with practically no inclination. The plains are common in the Atlantic and Indian Oceans. *Abyssal hills*, also common, rise from the plains with a maximum relief of 600 meters.

Oceanic rises are large isolated, generally nonseismic areas which rise to greater heights above the floor. The most famous of these is the Bermuda Rise whose volcanic Bermuda Pedestal is crowned with coral. Bermuda is the farthest land from the Equator in any ocean where living coral reefs thrive.

Seamounts abound in the open ocean with new ones being discovered each year. They are large underwater peaks—volcanic in nature—rising to very near the ocean's surface in many cases. They present navigational hazards to submarines, particularly to the later generations of deep-diving hulls. *Guyots* are flattopped seamounts generally deeper than 200 meters whose average depths are 1,000 meters from the surface. Guyots were unknown until World War II, when ships discovered these volcanic wonders by taking continuous fathometer traces. The guyots are mostly in the North and Central Pacific and the Mariana Basins.

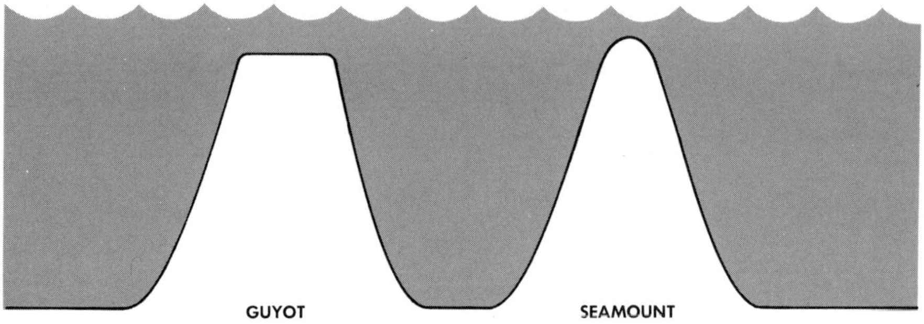

Figure 2-7 A Guyot and a Seamount

210 Mid-Ocean Mountain Ranges. The well-known Mid-Atlantic Ridge was discovered during the Challenger Expedition in 1873. Sounding lead on hemp line served to provide the primary indications. A more complete detail was provided in the period 1925–1927 by the German research ship, *Meteor*, with echo soundings. In 1929 a rise, now called the East Pacific Rise, was discovered. With this discovery there began a long line of investigations which revealed a distinct oceanic ridge and rise system. In fact, oceanographers have collected soundings which show a well-developed ridge in all major basins except the North Pacific. In the North Pacific an intermittent ridge lies in the west, while a longitudinal rise splits off dividing the Northeast Pacific Basin.

The Mid-Atlantic section of the great ocean ridge system occupies about 30 per cent of the entire Atlantic Ocean area. Island outcroppings are Iceland, Azores, Ascension, and St. Helena among others. While the other ocean basins have continuously developed mountain systems, they do not have outcropping above sea level to compare with the Atlantic high points.

Accumulated data have demonstrated that the continuous mountain range extends 40,000 miles throughout the basins in widths from a few hundred to a thousand miles.

Connected with the northeast Pacific Rise are several fracture zones. The fractures or faults extend from the North American coast for thousands of miles with average widths of 60 miles. To the north the Mendocino Zone appears to be connected with the San Andreas fault in California. The Murray fracture zone appears to be an extension of the east-west Tehachapi Mountains, north of Santa Barbara. The Clarion and Clipperton zones to the south, also running east-west, have not yet been fully explored. Similar east-west zones in the Atlantic recently have been discovered but little is known of them at this time. Also, recent oceanographic exploration has indicated the existence of well-defined fracture zones in the Indian Ocean.

211 The Atlantic Rift Valley. During the 1950s the Rift Valley, also called *cleft* or *graben* (German: ditch), extending along the crest of the Mid-Atlantic Ridge, was discovered. Physiographic presentations prepared then and extended in detail show that the Rift Valley averages 1,280 to 3,840 meters (4,200 to 12,600 feet) below sea level. Mountains rise an average of 1,820 meters (6,000 feet) above the valley floor, which in turn averages about 5,000 meters (16,400 feet) in depth. In comparative terms, the Mid-Atlantic Rift Valley is twice as wide, twice as deep, and much longer than the Grand Canyon.

The Rift Valley has been connected to the shallow oceanic earthquakes by its proximity, indicating that it may be a continuous north-south fracture zone. Scientific explorations have uncovered evidences of a worldwide rift system, suggesting many avenues of research concerning its origin, present state, and future development.

212 Bottom Sampling Instruments. The three basic types of bottom sampling devices are *corers*, *grabs* (or *snappers*), and *dredges*.

Figure 2-8 Bottom Samplers: Corer, Snapper, and Orange-Peel Grab

Corers are designed to obtain undisturbed, sediment samples called *cores* to depths as much as 70 feet below the ocean bottom. The deep* cores which are successfully retrieved average between 40 and 50 feet in length. There are corers which are designed to penetrate to depths of 10 feet or less. The latter would be used when the sediments are believed to have a *shallow thickness* or when the time element is important. Shallow corers normally are retrieved much more quickly than deep corers.

Grabs are utilized to sample loose surface material. A typical example is the orange peel design with its four jaws which resemble its namesake when closed. The *clamshell snapper*, a closely related instrument, also retrieves bottom surface material with its two clamshell-like jaws. Both the devices have the disadvantage of disrupting the material as the jaws are closed.

The last instrument, the *dredge*, is simply a tow device. It can be towed at speeds of two or three knots and is primarily used for scraping representative rock samples from the ocean floor.

213 Introduction to Sediments. The study of bottom sediments is important from many standpoints. Sediments have accumulated through the ages to smooth over the rough topographic features of the abyssal plains and have, until recently, led to false conclusions about many bottom areas. Investigation of deep core samples has repeatedly provided valuable clues to the earth's climatic and geologic history. Future generations, finding ashes in cores from the area of the North Atlantic trade routes, may not realize the ash accumulations were caused by nineteenth-century coal-burning ships. Many sediments have already gained some commercial value, especially the directly-mined manganese nodules and diamond debris. The sedimentation investigations have contributed to other associated oceanographic studies such as currents, biology, and air-ocean interactions.

The most systematic classification of underwater sediments is by origin. They have four fundamental beginnings called: *terrigenous*, *biogenous*, *halmyrogenous*, and *cosmogenous*.

214 Terrigenous Sediments. The *terrigenous* classification includes those inorganic sediments which arrive from the continents and are of a clastic (chemical or mechanical destruction) or volcanic origin. They have arrived in the form of fine dust clouds from great desert areas or have been brought from the shore areas by turbidity and surface currents. In fact, turbidity currents are considered to have been the primary carriers of the coarse terrigenous materials found in much of the abyssal depths.

* The corer classifications, deep and shallow, refer to sediment thickness rather than water depth.

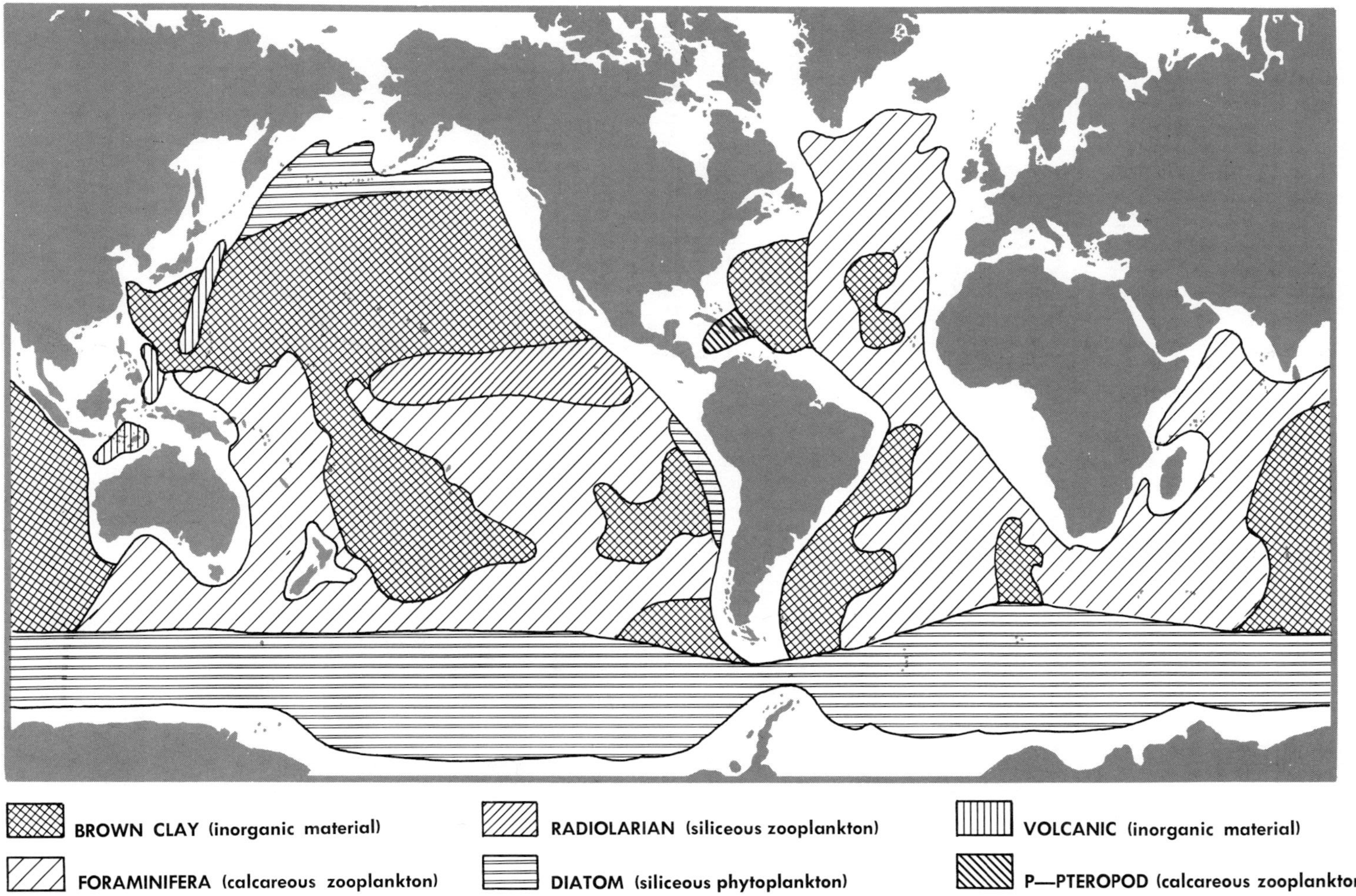

Figure 2-9 Sediments of the World's Oceans

The Sahara Desert has contributed to the bottom material west of the Cape Verde Islands in the Atlantic; regions near New Zealand also have been covered by the Australian Desert sands. This fine particulate matter has contributed largely to the widely distributed sediment now called *brown clay*. Its classical name had been *red clay* but the adjective *brown* describes much better its predominant color. Coarse and ungraded materials called *glacial marine sediments* are attributed to icebergs melting in deep water and releasing the moraine* type pebbles to the bottom. Subaerial and submarine eruptions have provided dust, which has blown or otherwise scattered, sinking and becoming *volcanic* terrigenous material.

Note in Figure 2-9 the entire North Pacific basin is covered primarily by brown clay. Other areas where brown clay is identified are the center section of the South Pacific, western North and South Atlantic, and the East Indian basins. Volcanic mud is predominant only west of New Guinea, an area where there have recently been volcanic disturbances.

215 Biogenous Sediments (Oozes).† The *biogenous sediments* have an organic beginning and are either *benthonic* or *planktonic*. A very small percentage of the bottom residue probably has a nektonic (free-swimming) origin; however, this small amount is not usually identified as such on the ocean floor. *Benthonic* sediments are decomposed remains, usually shells and frustules, of bottom dwellers, both plants and animals. The vast majority of the benthonic sediments are in the coastal zones where conditions are more favorable for life. Below the light zone, however, the benthonic organisms are few—there are no plants—and sediments of this type are therefore sparse.

Since those biogenous sediments having an origin in planktonic organisms are formed in the deep sea, they are usually termed *pelagic sediments*. Pelagic sediments, calcareous‡ and siliceous,§ are present in all regions of the ocean basins. This measurable material represents only a small percentage of the dying organisms in the oceans because most of the residue breaks up and dissolves before reaching the bottom.

The calcareous sediments or calcareous oozes, as they are called, are primarily from the shells of the globigerina, pteropod and coccolithophores. They are located in large areas of the South Pacific, the eastern North and South Atlantic and western Indian Ocean basins. The siliceous oozes generally represent the remains of diatoms and radiolarians. The phytoplankton, diatom, has its ooze around the Antarctic

* Rounded and unsorted.

† See Chapter 14 for complete discussion of marine biological terms.

‡ Calcium carbonate base.

§ Silicate base.

Continent, showing its affinity for cold waters; traces of the diatom oozes* have been found also in areas of upwelling. The zooplankton, radiolaria, prefers the warm regions as is evidenced by the belt of its sediments in the tropical Pacific.

It should be noted that decalcification of organic calcareous sediments occurs at depths greater than 4,000 meters. The hydrostatic pressure causes the dissociation of carbon dioxide and increases the solubility of calcium carbonate. Consequently, the calcareous shells at great depths become unidentifiable. Since the organic components of the shells are now dissolved, the remaining inorganic constituents become components of brown clay, previously discussed. It is possible that extensive future chemical analysis will provide enough evidence to allow some of the brown clays to be classified as biogenous. Certainly, some areas of these clay deposits are nonclastic and nonvolcanic in origin. Note that the brown clays seems to dominate in the greater depths (Figure 2-9).

216 Halmyrogenous and Cosmogenous Sediments. The halmyrogenous (Greek, *hal*-sea, *myron*-unguent) sediments are formed directly by chemical precipitation when the water becomes oversaturated. These authigenic (Greek, *authi* and *genes*, born on the spot) materials may build in layers to boulder size around a seed pebble or even a shark's tooth. Manganese and iron nodules are such examples, both of which have commercial value but neither is present in sufficient quantity to be important. The calcareous deposits in the Bahama Banks are also examples of authigenic formations.

The *cosmogenous* sediments, meteorite remains, are extraterrestrial in origin. They are quantitatively very minor, but do represent an ever increasing measure of the earth's mass. The tiny spheres average 0.2 mm in diameter and are usually magnetic iron or siliceous substances.

217 General Bottom Considerations. While at first glance the ocean bottom may seem to be a surface of unconcern to the mariner, there are some immediate bases for becoming familiar with simple geologic concepts. The nature of the solid bottom and its overlying sediments will influence the accuracy of sonic depth equipment and can adversely effect, by chemical action, operation of communication cables, oceanographic instruments, and other manmade devices. Shifting sedimentary materials can have destructive effects on pilings and offshore foundations—especially when the deep and bottom currents have not been observed and taken into consideration by the engineers. Bottom-bounce sonar is

* Diatomaceous earth (found in dried ancient sea beds) used in marine boiler insulation and filter systems is derived from diatom ooze.

able to perform much better when the sound rays are not absorbed by mushy or loosely packed sediments or scattered by rough bottom configurations.

If the cargo submarine is ever going to become a vessel of major commercial interest, it will be necessary to have accurate knowledge of the gross bottom topography for navigational purposes. In addition, it is often desirable to be aware of the nature of the bottom sediments, especially in shallow waters where the bottom is frequently dredged for shellfish and where ships frequently anchor.

218 Deep Submergence Vehicles. Future geologic exploration of ocean depths will be made by men working in those depths. This realization has drawn scientists from their desks to deep submergence vehicles, especially since Jacques Piccard and Lieutenant Don Walsh explored the Challenger Deep in 1960 in the *Trieste I* and reached a record depth of 35,805 feet. The first significant deep dive was made in the bathysphere occupied by Otis Barton and William Beebe to 923 meters (3,028 feet) on 15 August 1934. Since that time, many advances have been made—some directly resulting from the loss of the submarine *Thresher* on 10 April 1963.

Trieste I, developed by Auguste and Jacques Piccard, was purchased by the U.S. Naval Electronics Laboratory (NEL) in 1958. This bathyscaph, really a steel-hulled gasoline-filled balloon, had little horizontal movement capability and could remain submerged for only a short time. An improved model, *Trieste II*, was in service in time to assist significantly in the *Thresher* search and has supported investigations of acoustics, physical oceanography, and marine geology off the southern coast of California.

Submersibles are generally of three basic types: bathyspheres, bathyscaphs, and small conventional submarines. Bathyspheres are simply watertight spheres that can be lowered to any depth of interest by means of a cable. Their only propulsion capability in any direction is that afforded by the motion of the suspension cable.

A bathyscaph is a large gasoline-filled chamber and is therefore buoyant. In concept, a bathyscaph operating in water is identical to a dirigible operating in air. Because of its relatively large size, it is rather slow and maneuvers poorly, but it is undoubtedly the safest of the submersibles. Any loss of power results in ballast release and in the inherently buoyant vessel automatically coming to the surface.

A small conventional submarine consists of one or more watertight chambers, usually spherical or cylindrical. When, as is often the case, the propulsion and power systems are outside the pressure chamber, the vehicle can be quite small, since the greatest problem in constructing any type of submersible is to provide a personnel chamber capable of withstanding the tremendous pressures experienced in the deep sea (about 1,300,000 pounds per square foot at a depth of 20,000 feet, for example). Since somewhat more than 99% of the ocean is shallower than 20,000 feet, there seems very little need for a submersible with a capability greater than this. Just about all the development work on deep submergence vehicles seems to be concentrated on small submarines with a depth capability up to 20,000 feet.

Additional Reading

Bascom, Willard, *Waves and Beaches*, Doubleday and Co., Inc., 1964, (paper bound)

Bullard, E., "The Origin of the Oceans," *Scientific American*, September, 1969.

Bullen, K. E., "The Interior of the Earth," *Scientific American*, September, 1955.

Dunbar, C. O., *The Earth*, World Publishing Company, 1966.

Emery, K. O., "The Continental Shelves," *Scientific American*, September, 1969.

Ericson, D. B., Wollin, G., *The Ever-Changing Sea*, Alfred A. Knopf, 1967.

Fisher, R. L., Revelle, R., "The Trenches of the Pacific," *Scientific American*, November, 1955.

Heezen, Bruce C., "The Origins of Submarine Canyons," *Scientific American*, August, 1956.

Hurley, P. M., *How Old is the Earth?* Anchor Books, Doubleday & Company, Inc., 1959.

——, "The Confirmation of Continental Drift," *Scientific American*, April, 1968.

Menard, H. W., "The Deep-Ocean Floor," *Scientific American*, September, 1969.

——, *Anatomy of an Expedition*, McGraw-Hill Book Co., 1969.

Shepard, Francis P., *The Earth Beneath the Sea*, The Johns Hopkins Press, Baltimore, 1959.

——, *Submarine Geology*, 2nd ed., Harper and Row, Inc., New York, 1963.

Stetson, H. C., "The Continental Shelf," *Scientific American*, March, 1955.

Sweeney, James B., *A Pictorial History of Oceanographic Submersibles*, Crown Publishers, Inc., New York, 1970.

Turekian, K. K., *Oceans*, Prentice-Hall, Inc., 1968.

Urey, H. C., "The Origin of the Earth," *Scientific American*, October, 1952.

CHAPTER THREE

Water: The Common Denominator

301 Importance. The study of the environment requires that each of the components be studied in proportion to its importance or impact on the subject. Were it not for the water in our surroundings, commerce and trade would be impossible. More essentially, defense, weather, and living functions would be inexplicably modified. But even more basically, life itself would probably not have taken place. In the early dawn of time, this warm hospitable fluid offered a haven for the tiniest particle of circumstance that grew to be recognizable life on this planet. Just how nonliving things became living is yet unknown, and the production of the mysterious substance called protoplasm has yet to be fully understood. The importance of water cannot be overemphasized. If it were not for the presence of water vapor in the atmosphere and water filling the ocean basin, this planet would be uninhabitable due to temperature extremes alone. It will be demonstrated later how these extremes are ameliorated. Although the quantity of water in the forms of

Saturation diving with submersible decompression chamber
Courtesy: Ted Cox, Pasadena, Maryland

vapor and fresh water is slight, it is nonetheless a major factor in our environment. Ninety-eight percent of the mass of all existing water is seawater. It is estimated that if all the water vapor in the atmosphere were released in the form of rainfall, it would raise the level of the oceans only a single inch.

302 The Hydrologic Cycle. The interaction of the sea and the atmosphere is shown in a process called the *hydrologic cycle*. Large-scale evaporation takes place, as would be imagined, over the source of water, the ocean. This water, in vapor form, free from salt, is then carried by the winds over the land. Condensation releases the vapor in the form of rain, and the water finds its way back to the sea as continental runoff, principally from rivers.

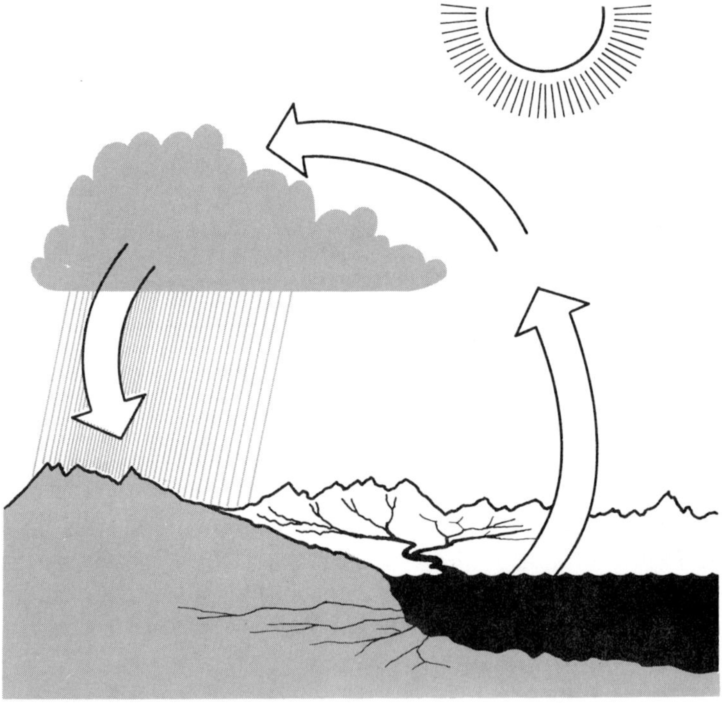

Figure 3-1 Hydrologic Cycle

One of the first significant properties of water is demonstrated in this cycle. Water has been called the universal solvent. This may be an exaggeration but water does have the ability to dissolve more substances, in greater quantities than any other liquid. Since in evaporation, only fresh water is removed from the sea, the fresh-water rain begins the leaching or rinsing action of the land. Although river water is normally thought of as "fresh," it is quite different from distilled or "pure" fresh water. In fact it is the slight amount of dissolved material which gives river water its distinctive and palatable taste. This rinsing action is partially the cause of the saltiness of the seas. Various theories are prevalent but there is little doubt that, in combination with volcanic activity, the salinity of the oceans is partly due to the dissolving of salts from the land, and their retention at sea by the selective process of evaporation.

303 Thermal Properties. Water is a truly amazing compound; for example, its surface tension is the highest of all liquids, mercury excepted. It exhibits strange properties that seem to be uncharacteristic of its apparent structure. If the water molecule (H_2O), consisting of two atoms of hydrogen and one of oxygen, followed the normal behavior of molecules of similar structure, such as hydrogen sulfide, hydrogen selenite, and hydrogen telluride, it would be expected to boil

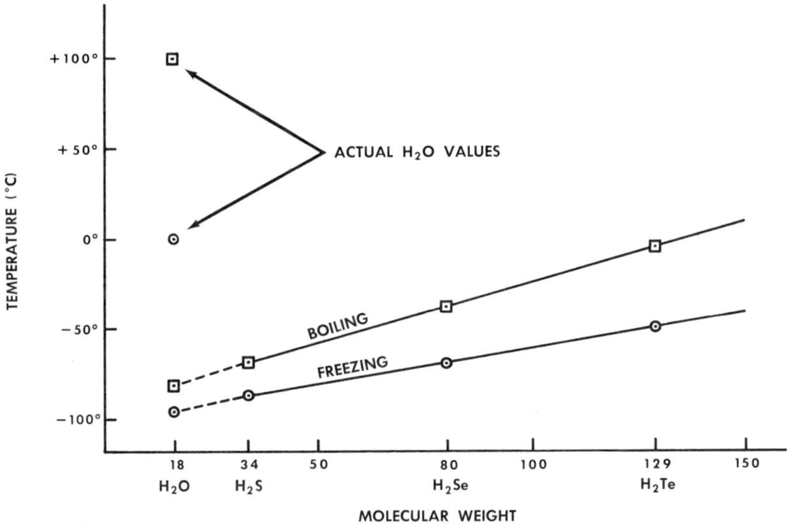

Figure 3-2 Boiling and Freezing Behavior of Similarly Structured Molecules

at −80 degrees C (Celsius) and freeze at approximately −95 degrees Celsius. Figure 3-2 shows this relationship graphically. The reason for this peculiar behavior is due to the unique structure of the water molecule itself. The two atoms of hydrogen arrange themselves in such a manner that they form an angle of 105 degrees with the oxygen atom. This gives the molecule polarity or an electrical charge orientation. Adjacent molecules are attracted and may cluster together in groups of several molecules united by this *hydrogen bond*, as it is called. This phenomenon is called *polymerization.* The tendency of water molecules to form these hydrogen bonds increases with decreasing temperature.

Fresh water will exhibit its most dense structure at 4°C (39.2°F). Further cooling produces more clustered molecules and a decrease in the density of the liquid. Finally, at the freezing point, 0°C (32°F), the decrease in density becomes most pronounced. Ice will increase its volume by ten percent over the volume of the original water.

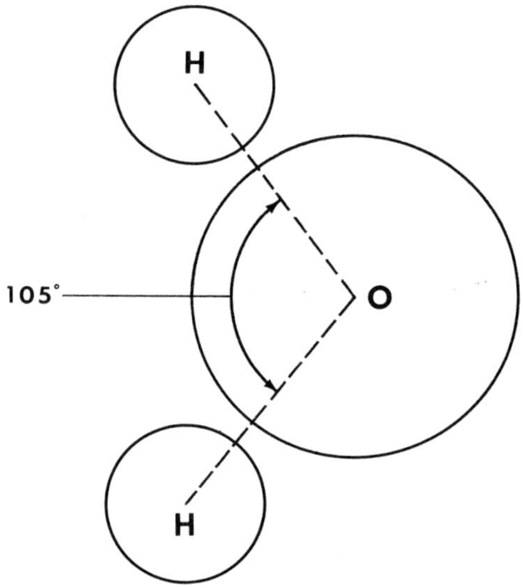

Figure 3-3 Structure of a Water Molecule (H_2O)

Water exists in three states: *solid*, *liquid*, and *gaseous*—under conditions of naturally occurring temperatures—although it is most abundant in liquid form. Due to its abundance, water has been selected as a convenient standard for heat measurement.

At this point it appears necessary to clearly define the two concepts of *heat* and *temperature.* A concept of temperature is normally gained early in life as a bodily sensation. An object is either hot or cold. Later an individual gets the concept of temperature from observing the thermometer. Since the activity of molecules in the structure of a gas reflects the amount of kinetic energy present, this factor is directly related to heat and temperature. A high temperature reflects a high average molecular kinetic energy, and a low temperature reflects a low average molecular kinetic energy. A simple analogy can be drawn, showing an individual sounding a tank with a stick. He is measuring the average level in the tank, i.e., $\frac{1}{2}$ full, $\frac{3}{4}$ full. This would be analogous to inserting a thermometer in a liquid to determine the average kinetic energy of the molecules. In sounding the tank, no information as to the quantity of the contents would be gained unless the dimensions of the tank were known. Similarly the total quantity of heat cannot be determined by simply taking the temperature of a substance, unless additional information is provided.

The same amount of heat can be added to two different objects at the same initial temperature, yet the resulting temperature of the two can be vastly different. For example, consider the equal masses of water and aluminum at room temperature. After being subjected to heat from identical bunsen burners for the same period of time, thereby adding the same quantity of heat, the resulting temperatures of the two will be quite dissimilar. This result indicates that water and aluminum have different specific heat capacities.

The ability of water to absorb heat is very great. In fact it has the largest capacity for heat of all solids and liquids with the exception of ammonia (NH_3). Because of this, water has been chosen as the standard for the measurement of *specific heat capacity*. Specific heat capacity of any substance is defined as the quantity of heat required to raise one gram of that substance one degree Celsius (at a constant pressure). Stated another way the specific heat capacity of water is one calorie per gram per degree Celsius.

Recalling the unique structure of the water molecule, the change of state of water will now be discussed. As said, water can exist in three states; solid, liquid, and a gas. Twenty calories of heat applied to a gram of water at 80° Celsius, will raise its temperature to 100° Celsius. However, the application of one more calorie of heat will not raise it to 101°. Since the boiling point of water is at 100° Celsius, it must now be supplied with sufficient heat energy to break the hydrogen bonds and change the state of the compound to the gaseous state, at the same temperature. At the boiling point it is estimated that 80 percent of the hydrogen bonds remain. The number of calories required to make this change of state for one gram of water is approximately 540—over five times the heat required to raise the same gram of water from freezing to boiling. The heat required to vaporize is greater for water than any other substance. This quantity of heat, necessary to change state without changing temperature, is called the *latent heat of vaporization.*

This is an extremely significant factor in the world heat budget because it allows the transfer of enormous quantities of heat over the earth's surface. Each gram of water which changes state from liquid to gas takes up this quantity of heat. It is little wonder that, as a swimmer emerges from the swimming pool into a breeze, evaporation (the change of water from liquid to gas) removes great quantities of heat from the surface of his body, causing great caloric loss at the skin, the equivalent of a tremendous drop in air temperature. This latent, or "stored," heat will naturally be released when the water vapor condenses and falls as rain. In this instance the heat is released, not to the rain, but to the surrounding air.

The quantity of water vapor present in the atmosphere is always of vital importance to the weather forecaster, because water vapor produces most weather phenomena. Clouds, rain, snow, frost and dew, are the natural manifestations of the quantity of water vapor in the air. Note here that water vapor is invisible. To many, water vapor means visible steam. This is very far from the case. A close look at a boiling tea kettle will disclose a small transparent space between the spout and the outpouring steam. In this space where nothing appears to be, is water in true vapor form. The steam that is so evident, is visible moisture which has condensed from the vapor form upon contact with the relatively cold room air. This is much the same process that forms visible drops of moisture when air comes in contact with the cool side of an iced drink in summertime.

Figure 3-4 Condensation

304 Humidity. In measuring this most vital constituent of the atmosphere certain standards are used. *Absolute humidity* is an expression which defines the total mass of water vapor present within a given volume of air. The units of this measurement are grams per cubic meter.

Probably the most common term used in relating water vapor content is *relative humidity*. Simply stated, relative humidity expresses the quantity of water vapor present in the atmosphere as compared to the quantity that the existing air could hold at its ambient temperature. This ratio is always expressed as a percentage. A relative humidity of 80 percent means that the air is within 20 percent of holding all the water its present ambient temperature would allow. In the early 19th century, John Dalton published his famous law concerning the partial pressures of gas constituents. He showed that if a gaseous mixture exhibited a certain pressure against the confines of a container, this pressure was made up of the total of all the individual pressures of the component gases in the mixture. Experiments show that, in a container, partially filled with water and with the remainder filled with air at the same temperature, evaporation will begin to take place. As the molecules of water enter the atmosphere as a gas, the pressure in the air will begin to rise, because of the individual contribution to the pressure made by the water vapor, as Dalton predicted. When the number of molecules evaporating equals the number of molecules condensing, a state of equilibrium exists. The pressure measured at that time is called the *saturation vapor pressure* (e_s) of the liquid and 100 percent relative humidity exists. The formula for relative humidity is:

$$R.H. = \frac{e}{e_s} \times 100 \qquad (1)$$

where e is the existing vapor pressure.

In practice, the relative humidity is usually computed as based on temperatures measured by a sling psychrometer (Figure 3-5), and the form (2) of the equation is used to determine (e)

$$e = \frac{R.H.(e_s)}{100} \qquad (2)$$

and other humidity parameters of interest to the meteorologist; (e_s) is available in tabulated form for any given temperature.

This concept of vapor pressure is also important, in Chapter 15, because it concerns the formation of certain weather elements. It is important to note that the capacity for air to hold water vapor depends on temperature. The absolute humidity will not change unless the air's capacity to hold water vapor is reached and some water vapor is removed by condensation. On the other hand, the relative humidity will change as soon as the air's ability to hold water vapor (i.e., the temperature) changes at all. No change in the absolute humidity need accompany this change in relative humidity.

Another term which requires explanation in dealing with humidity is the *dew point*. The ability of air to hold moisture is dependent upon the temperature. The warmer the air becomes, the more water vapor it can hold. As it becomes colder, the capacity for holding water becomes less. As temperature decreases, the relative humidity increases until it is 100 percent. At this point the air is saturated and the dew point temperature has been reached. Simply, this is the temperature where dew (condensation) will form.

To assist the meteorologist there are instruments with which he can measure the amount of water vapor present. The search for a material that would respond to changes of moisture content gave unexpected results. Oddly, one of the best measuring devices of humidity is human hair. Most people notice the change in the behavior of their hair as the weather begins to become exceedingly dry. The moisture present in the air varies the distance between the cells of the hair and lengthens or shortens it accordingly. This movement may be transmitted and magnified through a series of cranks and levers and displayed on a graduated scale to indicate the humidity; such an instrument is called the *hygrometer*. If the movements of the indicator are recorded by the use of a pen tip, on a rotating drum, the instrument is called the *hygrograph*.

As a general rule, an instrument that gives only a single discrete reading of a factor is called a ____ *meter*, e.g., *thermometer*, *barometer*, while one that gives a time record of the variation of a factor is called a ____ *graph*, e.g., *thermograph*, *barograph*.

Another device for determining humidity is the *sling psychrometer*. It consists of two thermometers mounted side by side on a board attached

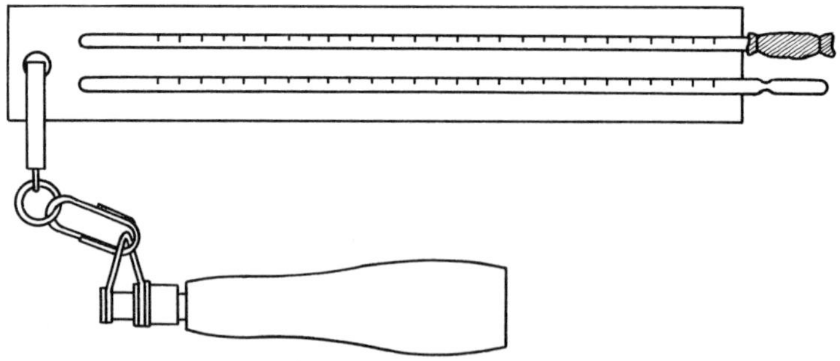

Figure 3-5 The Sling Psychrometer

to a short handle by a few links of chain. This allows the device to be swung, or extensively ventilated, which is essential to obtain accurate readings. The bulb of one of the thermometers is covered by a small muslin cap dampened with water when a reading is to be taken. When whirled about, the uncovered thermometer will read the ambient temperature. The other thermometer will be cooled by evaporation to a degree dependent upon the present saturation of the air. The closer the ambient air is to saturation, the less effect the evaporation will have upon the other thermometer. These two thermometers are referred to as the *wet bulb* and the *dry bulb*. The difference between these two readings can be converted to relative humidity through the use of precomputed tables.

It is important to understand the distinction between wet bulb temperature and dew point temperature. The *dew point* can be produced by cooling the air to its saturation point and then recording that temperature, whereas the *wet bulb temperature* is the lowest temperature obtainable when cooling the air by evaporation to saturation. When the relative humidity is 100 percent, the wet bulb and the dry bulb readings are the same. In this case the wet bulb reading could not be lowered below the dry bulb by evaporation, because at conditions of 100 percent relative humidity no evaporation could take place. There will never be a case when the wet bulb temperature exceeds the dry bulb.

A third method of measuring moisture is routinely used to take soundings at various altitudes or to determine the humidity variation with height. This is the use of lithium chloride installed in the *radiosonde*. This device will be covered in detail in Chapter 4.

Each of the aforementioned devices has limitations. The sling psychrometer has an obvious limitation when the temperature is such that freezing might take place at the wet bulb. The hair hygrometer and the radiosonde are very unreliable in the region of low humidity readings. In short, although these devices are serviceable, considerable improvement could be made in humidity measuring devices.

305 The Solid Form. At the other end of water's temperature spectrum, its solid state, *ice*, is encountered. In a fashion similar to the preceding discussion when a gram of water at 20°C is divested of 20 calories of heat, its resulting temperature will be 0°C in the liquid state. The removal of another single calorie from this gram of water will not lower the temperature to −1° C, since the phenomenon of latent heat is again encountered. Approximately 90 percent of the hydrogen bonds still remain at the freezing temperature. It is necessary to remove a significant quantity of heat to allow freezing. The same gram of water at 0°C must have approximately 80 calories of heat removed before ice will form at the same temperature. The quantity of 80 calories is called the *latent heat of fusion*. It is apparent that this is nearly as much as is necessary to raise the temperature of a gram from the freezing point to the boiling point. Water's latent heat of fusion is the highest of all substances except ammonia.

The fact that water does not exhibit its most dense structure as a solid again puts it in a unique position. This occurs with only three other substances in nature; bismuth, antimony and gallium. This of course, has a profound effect on biological life in fresh water lakes and rivers. The fact that ice is less dense than liquid water, and consequently floats, allows life in fresh water lakes and rivers to be protected from severe cold. The recent interest in delaying human life through cryogenics (freezing bodies in a state of suspended animation) has been stymied by this fact that any organ or tissue (with minor exceptions) is destroyed by the expansion of its water content as it freezes. Another interesting aspect of ice's thermal expansion is that the weathering of continents has been made possible through this medium. Water soaks into the crevices, caves and fissures in mountain and rock surfaces. A drop in temperature turns this water to ice, and the resultant expansion exerts tremendous pressures—sufficient to cause fracture of the rock itself. Anyone who has been unfortunate enough to be caught without anti-freeze in his automobile's cooling system can also sadly attest to the ability of expanding ice to break extremely strong materials.

306 Seawater. The discussion up to this point has dealt mainly with fresh water. However, 98 percent of all water is seawater. In 1872 HMS *Challenger* sailed from Spithead, England, on an oceanographic exploration. Until this time the depths of the oceans had been sparsely sounded and investigated. Upon her return in 1876, she brought back voluminous information on all the seas of the world with the exception

Table 3-1 Dittmar's Constituents of Seawater

	PARTS PER THOUSAND (‰)	
Chlorine	18.980	
Sodium	10.561	
Magnesium	1.272	
Sulfur	0.884	
Calcium	0.400	(Cl = 19‰)
Potassium	0.380	
Bromine	0.065	
Carbon	0.028	
Strontium	0.013	
Boron	0.004	

of the Arctic. The data were major contributions to the fund of man's knowledge in fields such as meteorology, hydrography, geology, botany, the chemistry and physics of seawater, petrology, zoology, and geography. The water samples brought back from the far reaches of the world's oceans were analyzed by Dittmar, and the major constituents are found in Table 3-1.

The major constituents are roughly defined as those whose concentration exceeds .001 parts per thousand. In addition to those already listed, there are minor constituents which bring the total number of elements found in seawater to over 50. Dittmar's work has brought about the concept, "Law of relative proportions," which states regardless of the absolute concentrations of total dissolved solids, the ratios between the more abundant constituents are virtually constant. This "law" has been useful in the study of ocean waters but breaks down when restricted waters are considered, or extreme discretionary accuracies are demanded. The application of this principle requires open ocean conditions. The composition of the ocean waters is important because the dissolution of substances in water changes various characteristics of water in the pure state.

When one speaks of seawater, the term "salty" is often used. To many this term means table salt or sodium chloride (NaCl) but when the oceanographer speaks of salinity the term is more encompassing. *Salinity* ($S°/_{oo}$) in the sea refers to the total amount of solid material dissolved in a kilogram of seawater when all of the carbonate has been converted to oxide, all bromine and iodine replaced by chlorine, and all organic matter completely oxidized. It is usually expressed in grams per kilogram or parts per thousand ($°/_{oo}$). Determining salinity by this definition is extremely cumbersome, so that another parameter called chlorinity ($Cl°/_{oo}$) is defined. *Chlorinity* is essentially defined in terms of the total halogen content in grams per kilogram.

What is the effect of all these considerations? How does one find the concentration of the various constituents in a seawater sample? Salinity has been defined. The "law" of relative proportions has been established. Chlorinity has been specified and is capable of being chemically separated. What now? Near the turn of the twentieth century Knudsen determined that an empirical relation existed between salinity and chlorinity. This was updated in 1967 and is now expressed as follows:

$$\begin{aligned} \text{salinity} &= 1.8066 \times \text{chlorinity} \\ S°/_{oo} &= 1.8066\ Cl°/_{oo} \end{aligned} \qquad (3)$$

This has been adopted internationally and has provided the key to the major constituents, making it possible to take a sample of seawater, and by means of chemical titration, isolate the quantity of chlorinity, utilize the formula to determine salinity, and then apply the law of relative proportions to determine the actual value of all major constituents—a circuitous but practical method.

What of the effects of the dissolved solids? How does salinity affect the behavior of water? As a general statement, solutions, which include salt water, have properties which fresh water does not have. These are called the colligative (Latin—bound together) properties. They are *osmotic pressure*, *vapor pressure*, *freezing point depression* and *boiling point elevation*. The effect on vapor pressure and on boiling point are relatively slight, whereas the effect on freezing point and on osmotic pressure are quite large.

One result of the addition of salt is that the electrolytic properties are radically changed. Pure water has the highest dielectric constant of all liquids, making it a good insulator. The hazard in handling electrical appliances with wet hands is caused by the dissolved solids or impurities that make it a dangerous conductor. With the addition of more dissolved solids, water becomes a better conductor of electricity and the quantity of solids is directly proportional to the amount of conductivity increase. This unique behavior has led to the use of electrical conductivity for salinity measurements. In fact, this method is rapidly replacing the titration method as a determinant of salinity, due to increased accuracy and speed.

307 Man and Water. The unique properties of water touch the lives of all men. They bathe in it, sail on it, swim in it and drink it. Without it they would all perish. If this were not proof enough of the importance of this vital substance, consider that man's body is composed of about 80 percent water and the blood in his veins is 92 percent water with a striking chemical resemblance to the seawater. This firmly establishes water as the common denominator between man, the sea, and the atmosphere.

Additional Reading

Buswell, A. M., and Rodebush, W. H., "Water," *Scientific American*, April, 1956.

Byers, Horace, R., *General Meteorology*, 3d ed., McGraw-Hill, Inc., 1959.

Davis, Kenneth S., and Day, John A., *Water: The Mirror of Science*, Doubleday & Co., Inc., Garden City, N. Y., 1961.

Dietrich, Gunter, *General Oceanography*, John Wiley and Sons, New York, 1963.

Duxbury, A. C., *The Earth and Its Ocean*, Addison-Wesley Publishing Co., Reading, Mass., 1971.

Huschke, Ralph E., *Glossary of Meteorology*, American Meteorological Society, Boston, Mass., 1959.

Riehl, Herbert, *Introduction to the Atmosphere*, McGraw-Hill, 1965.

Sverdrup, H. V., Johnson, Martin W., and Fleming, Richard H., *The Oceans*, Prentice-Hall, Inc., 1942.

Williams, Jerome, *Oceanography*, Little, Brown and Company, Inc., Boston, 1962.

CHAPTER FOUR

Atmosphere: The Envelope

401 The Air. The preceding chapter has described that vital constituent of the atmosphere, water vapor, whence all weather comes, but what of the air itself? Each person goes about his daily living, breathing air in and out about eight million times a year, without giving it so much as a passing thought. Its importance for life need not be emphasized.

The atmosphere (Greek, *Atmos*—vapor) is the gigantic ocean of gaseous fluid which envelops the earth. It reaches into every hollow, every basin. It occupies almost every crack and crevice on the earth's surface and exists between the very pages of this textbook. From the surface of the earth upward, it encounters no boundary, but blends imperceptibly with the low density gas which occupies interplanetary space. This "vapor" sphere can be divided into a number of categories, depending upon the area of interest.

402 Constituents. The chemist is interested in the atmospheric chemical composition; the geophysicist is interested in its magnetic

A view of the envelope from the ionosphere

properties. Each man usually focuses his attention on a certain aspect or set of properties. However, the environmental scientist has an interdisciplinary outlook. He is concerned with all phases of the air, from its military applications to its dynamic behavior, to its influence and interaction with the sea.

The atmosphere can be divided into two broad categories, the *homosphere* and the *heterosphere*. The homosphere (Greek, *homo*—same) is the region where the atmosphere has a constancy of composition in much the same fashion as does the ocean. The major constituents of dry air, those which are of main concern, show no variation in their volumetric occurrence.

Table 4-1 Major Constituents of Dry Air

Nitrogen (N_2)	78.08%
Oxygen (O_2)	20.95%
Argon (Ar)	0.94%
Carbon dioxide (CO_2)	0.03%*

Other constituents are present, but they have no practical significance to the environmental picture since they are in such minute concentrations. For example neon, helium, krypton, and hydrogen are all present in much less than 0.00001 percent. Although oxygen is listed as diatomic oxygen, later discussion will show the occurrence of monatomic and triatomic oxygen. In fact, this triatomic oxygen (O_3) or ozone becomes a very significant part of the structure of the atmosphere.

Notice that these are referred to as the constituents of dry air. The term *dry air* is used in two different frames of reference. They are normally not confused since this will be the last time dry air is referred to in the sense of "air without any moisture content." In other discussions dry air is referred to as that "which has not reached its total capacity to hold water vapor."

Water vapor is a variable constituent of the air. It is present in amounts up to about four percent by volume. A global average of water vapor is approximately one percent. The molecular weight of water vapor is less than that of dry air, therefore it has less density. It then would appear that moisture would be concentrated in the higher levels of the air structure, but this is not the case. The ability of the atmosphere to hold water vapor is dependent upon temperature, as was discussed in the previous chapter. Since the temperature of the atmosphere decreases with altitude, the capacity to hold water decreases with it. In addition,

* There is some evidence that the CO_2 content has increased slightly over the past fifty years due to increasing use of hydro-carbon fuels.

the water sources, such as the oceans, rivers, and lakes, are all at the earth's surface. Therefore the greatest concentration of water vapor will be found in the lower layers of the atmosphere.

403 Temperature Structure. The temperature characteristics of the atmosphere are of vital concern to the meteorologist. The homosphere is the area above the earth which is made up of three regimes, as defined by temperature. They are:

	AVERAGE ALTITUDE
Troposphere	0– 7 miles
Stratosphere	7–30 miles
Mesosphere	30–50 miles

Since the air is a gas, it behaves according to the universal gas law. This law says that the absolute temperature of one gram-molecular weight of a gas is directly proportional to its volume and pressure. When a gas is pressurized, either the volume goes down or the temperature goes up, or both. Everyday experience shows this. When air is squeezed, in a tire pump, the temperature goes up. When a CO_2 fire extinguisher is activated, the gas expands and cools rapidly.

Used to probe the vertical structure of the atmosphere is a device called a *radiosonde*, which is a small (few pounds) instrument package of sensors attached to a radio transmitter that relays information back to a receiver on the ground. This radiosonde is carried aloft by a large balloon and will often reach altitudes of 100,000 feet. The instrument "fish" is considered expendable. The temperature and relative humidity information, which are sent back at standard pressure levels, may be converted to heights by the use of the gas laws. New equipment, using a *transponder* (a radio transmitter which operates only when triggered by an incoming signal), measures the time lapse of signal transmission to determine slant range. In conjunction with azimuth and elevation angle, very accurate heights can thus be obtained.

A *lithium chloride sensor* is used for the humidity information. Its electrical properties change with a change in moisture content of the air. The temperature element is a ceramic-coated element which also changes electrical resistance with a change in temperature. The pressure device is an aneroid element (Section 408).

To get wind information, the azimuth and elevation angles of the device are observed as the balloon ascends.

Solar heating is concentrated in equatorial regions. The volume of the atmosphere expands because of this heat, resulting in a thicker tropospheric layer (10 miles) than would otherwise be expected. Conversely, over the poles, the cooling is evidenced by a layer that is thinner than average (about 5 miles).

The tropospheric temperature normally decreases with height. Figure 4-1 shows the average temperature conditions at mid-latitudes for a standard day. This standard atmosphere has been adopted by international agreement and has been widely accepted by scientists. The surface temperature for the standard day is defined to be 59°F (15°C).

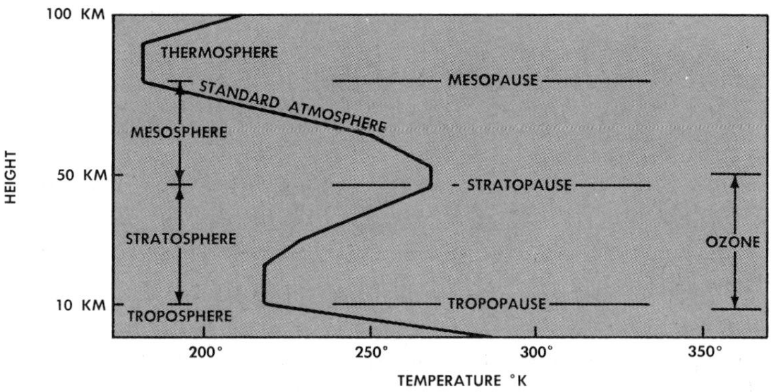

Figure 4-1 Temperature Structure of Atmosphere at Mid-Latitudes

Lapse rate refers to the way temperature changes with height. This term must have come into use when dealing with the troposphere, for it says "this is how the temperature 'lapses' as the altitude increases." Stated another way, a temperature which decreases with increasing altitude results in a positive lapse rate, since the lapse rate is defined as $-\Delta T/\Delta z$. Conversely, an increase of temperature with height is referred to as a *negative lapse rate* or *inversion*. At first, these terms may seem awkward, but with usage, they will become meaningful. A further discussion of lapse rate will be found in Chapter 7.

The variation of water vapor with height is such that, close to the surface, 1 percent is a reasonable figure, while at about 25,000 feet, it has decreased to 0.05 percent. Since water vapor is an essential determinant of weather, the troposphere is the true realm of most perceivable weather.

The character of the temperature lapse rate changes between the troposphere and the stratosphere. This is, in effect, the boundary between the two regions called the *tropopause*. There are similar "pauses" between the other layers in the atmosphere as determined by the temperature lapse rate. The stratosphere has a dual regime of temperatures.

The region between 7 and 12 miles is approximately isothermal. Above this isothermal layer up to the *stratopause* (transition layer above the stratosphere) the temperature gradually increases to near the freezing temperature of water.

The mesosphere is located above the stratopause between 30 and 50 miles in height. From the relative temperature peak at the stratopause, the mesosphere exhibits a region of positive temperature lapse rate (decrease with increasing height), to a point where the most severely cold temperatures in the atmosphere are found. This extremely cold layer called the *mesopause*, caps the mesosphere and forms the boundary between the mesosphere and the next adjoining layer, the thermosphere. Most of the thermosphere lies outside the homosphere and will be considered separately.

404 Chemosphere. There are approaches other than temperature to atmospheric investigation. Certain chemical processes take place which have given rise to the name *chemosphere*. Considered to extend from the tropopause to an altitude of about 120 miles, this layer shows considerable overlap of the homosphere and the heterosphere. The homosphere is of particular interest. The region of ozone production, designated the *ozonosphere*, is centered in the stratosphere and is responsible for the character of the lapse rate there. This most significant reaction, particularly as far as human life is concerned, is produced by the dissociation of molecular oxygen (O_2) into atomic oxygen and the continuous re-formation into ozone (O_3). This is a heat-releasing process. The ozone readily absorbs solar ultraviolet radiation; this dissociation-formation is a continuous process. The ultraviolet radiation absorbed is in such wavelengths that, were it not prevented from reaching the earth's surface, living organisms on the earth would be unable to survive. Ozone is the only constituent of the atmosphere capable of providing a shield in this narrow, but potentially lethal, segment of the electromagnetic spectrum.

Airglow (or *night glow* when it appears at night) also occurs in the chemosphere. Spectrographic analysis gives the best explanation for this effect. The glow may be the result of absorption of solar ultraviolet radiation by atomic oxygen, which is present in the spectrum. By comparing the angle of the elevation of this glow at several locations, estimates place it at a probable altitude of 60–100 miles. The airglow is difficult to study during daylight hours, but at night, it is clearly seen as a general spread of diffuse light across the sky. Recent observations by American astronauts confirm the occurrence of luminous haze in the altitude range where it had been thought to exist.

Another unusual aspect of this high-altitude region is the appearance of clouds that seem to glow after nightfall. They are called the *noctilucent* (Latin, *noctis*—night, *lucere*—to shine) *clouds*. These are normally observed at high latitudes during the summer season. Their height is roughly 50 miles and because of this extreme height, they direct reflected sunlight to an observer at the surface, long after the sun has disappeared. Rocket sampling of these clouds indicates that they are not composed of water vapor, but of cosmic dust with a high nickel content. There are surprising indications that these clouds may be lightly covered with ice, however. Water in any state existing at temperatures near $-100°C$ would be very unexpected indeed.

405 Heterosphere. Above the mesopause, at an altitude of 55–60 miles, the turbulent mixing of the atmosphere is reduced to the point that the composition of air is no longer constant. This is also the base of the heterosphere (Greek, *heteros*—different). In this region molecules and atoms tend to separate and arrange themselves in such a manner that the heavier ones predominate at lower levels while the lighter are found at extreme altitudes. The upper reaches of the heterosphere are called the *exosphere* (Greek, *exo*—outside). At these extreme heights, the atmospheric density is so low that a molecule with sufficient velocity can escape the gravitational attraction of the earth. This is possible because the low density gas allows the mean free path of an electrically neutral particle to be one mile at an altitude of 200 miles. At 500 miles height, a molecule will not encounter another one for 100 miles. Unless a particle collides with another particle, it can easily escape the earth's influence. This compensatory loss of mass may be the balancing effect that offsets the continuous gain in mass due to cosmic dust.

That area where the temperature increases with height marks the beginning of the *thermosphere*. The farther from the earth's surface, the less dense is the atmosphere. Above approximately 200 miles the familiar concept of temperature loses its significance, and direct radiation measurements from the sun are more meaningful.

406 Ionosphere. The source of all energy, the sun, provides another interesting effect in the upper atmosphere. This is *ionization*. When an electrically neutral atom gains or loses an electron, it becomes electrically charged. This can be produced by the absorption of solar radiation under certain circumstances. The study of this process in the upper atmosphere is called *aeronomy*. The well-known Kennelly-Heaviside layer has become a factor in long-range radio and radar performance. This layer led to further study and classification of these ionized regions. Scientific investigation has delineated four rather distinct regions of ionized particles, principally electrons. It has become standard practice to refer to them as the D, E, F_1, and F_2 regions. The A, B, C regions, previously thought to exist, did not verify under investigation. These regions are affected by the latitude, season, time of day, changes in solar radiation, and even by the longitude. The daytime heights above the surface of the earth are listed in the following table.

Table 4-2 Daytime Heights of Ionospheric Layers

REGION	MILES	KILOMETERS
D	35–55	60–85
E	55–90	85–140
F_1	90–150	150–250
F_2	150–600	250–1,000

The daily variation has been investigated and apparently the D region disappears at night. In addition, the distinct difference between the F_1 and F_2 regions also appears to diminish so that the two F regions merge as one. These changes raise the refracting layer for radio waves and logically explain the characteristically greater radio communications ranges at night.

It is obvious that, for tactical purposes, this skip distance, or long-range communication, can be either a serious handicap or a tremendous

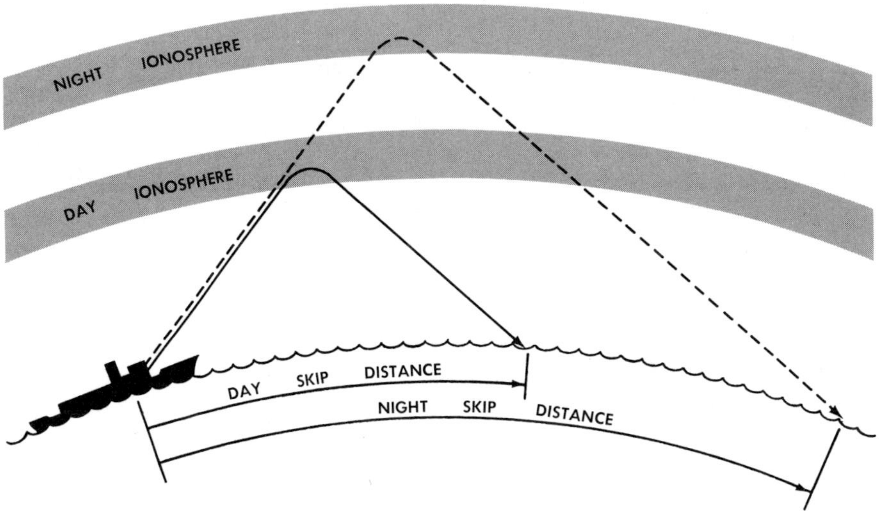

Figure 4-2 Electronic Skip Distance of Day and Night Ionosphere

advantage. The security of radio communications leaves a great deal to be desired under nighttime conditions. On the other hand, these same conditions are anticipated with great pleasure by amateur radio operators and short-wave enthusiasts.

407. Magnetosphere. Another characteristic of the earth that lends itself to study and classification is magnetism.

The field of a magnet is classically shown by the pattern arrangement that results from iron filings being acted upon by the magnetic forces. Similarly, the lines of the magnetic field of the earth arrange themselves in a regular pattern emanating from the magnetic poles of the earth. These lines would be found in a symmetrical fashion were it not for a force in space which distorts this pattern. The sun and its envelope of *solar wind*, also called *corona* or *plasma*, an outflow of high-speed protons and electrons, exert a pressure which compresses the boundary between the geomagnetic field and the solar plasma to a distance of ten earth radii (40,000 miles) on the sun (or day) side and to an undetermined distance (initial calculations put it in the vicinity of 50 earth radii) on the night side of the earth. The boundary between the earth's magnetic field and the solar plasma is called the *magnetopause*.

The space around the earth encompassed by the magnetopause has also been referred to as the *geomagnetic cavity*.

During periods of high solar activity it appears that radiation particles penetrate the magnetopause in a region where the boundary is not strong, that is, on the distant night side of the earth. These particles have arranged themselves, under the influence of the magnetic field of the earth, in two doughnut-shaped belts and one broad region of particles. These three divisions seem to be arranged according to relative charge strength and type of charge (i.e., electrons or protons). The study and classification of these trapped radiation areas were carried out principally by James Van Allen for whom they are named. Possibly these trapped particles may present a hazard to men in space.

An interesting and remarkable phenomenon associated with the magnetosphere is the *aurora*. Because this phenomenon has an appearance of approaching sunrise it was so named after the Roman goddess of dawn, Aurora. The name has been modified by the word *borealis* (Greek, *boreas*—north), or *australis* (Greek, *auster*—south). Aurora borealis has also been called northern lights. This effect is normally seen in a roughly circular band around the earth at about 65° to 70° magnetic latitude. Most recent theory maintains that, during periods of great solar activity, proton-electron plasma enters the magnetosphere in much the same way the Van Allen belts are formed. It is accelerated

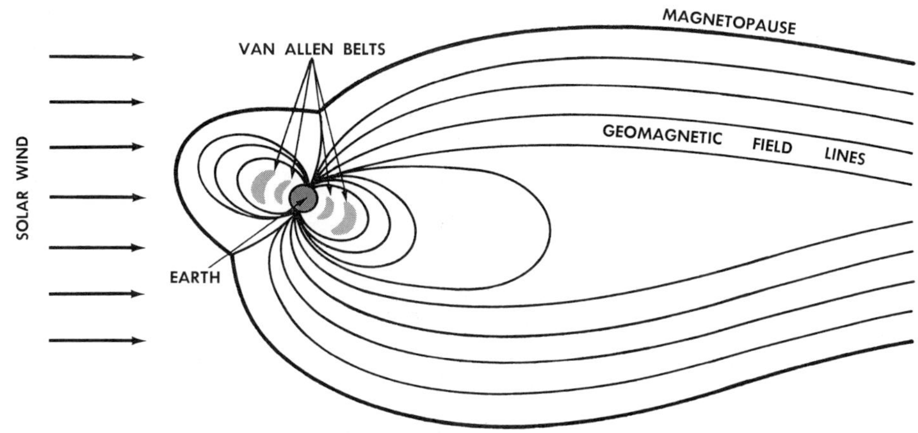

Figure 4-3 The Magnetosphere

towards the poles and oscillates between north and south along lines of the geomagnetic field. The auroras fall into two general categories, those that have a ray-shaped pattern, and those that do not. Green, greenish yellow, and greenish blue are the most common, but red, blue, orange, and violet have been seen during periods of exceptional solar activity.

408 Atmospheric Pressure. In 1643, one of Galileo's students, Torricelli, investigated the behavior of mercury in a glass tube. He took a glass tube full of mercury, about three feet long, closed at one end, and while he held the mercury in by putting his thumb over the open end, he inverted the tube and plunged it into a bowl of mercury. Carefully easing his thumb off the submerged end of the tube, so as not to allow any air to enter, he found that the level of the mercury in the tube dropped about six inches. It came to rest about 30 inches above the level of the mercury in the bowl. What was in the top of the tube now? A vacuum obviously, since nothing could enter. He reasoned, and correctly so, that the pressure of the air on the surface of the mercury in the bowl, had forced, or was holding, the mercury at a certain level in the glass tube. Torricelli had produced a vacuum, and invented the barometer as well. The basic barometer, as described, has had little refinement over the many centuries.

Actually the mercury column in the barometer adjusts its length until the pressure it exerts is exactly equal to the pressure exerted by the atmosphere on the mercury pool in the bowl.

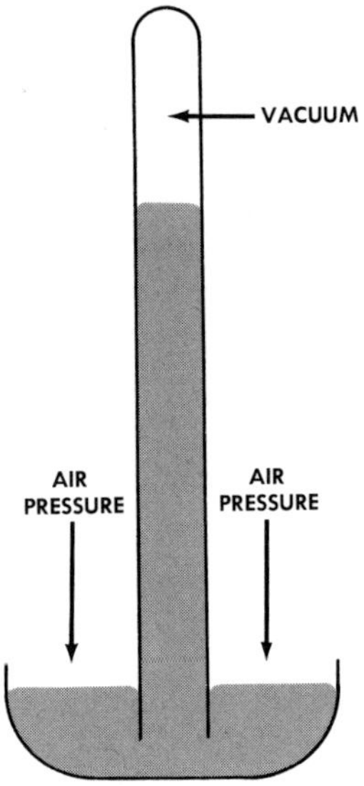

Figure 4-4 The Barometer in Cross-Section

Since pressure is force per unit area (force/area), the pressure exerted by the mercury column is simply its weight per unit volume multiplied by the column height, or:

$$\frac{\text{weight}}{\text{volume}} \times \text{height} = \frac{\text{weight}}{\text{area}} = \frac{\text{force}}{\text{area}} = \text{pressure.}$$

In addition, weight per unit volume is simply g times the density (ρg) so that the pressure exerted by a fluid column is given by:

$$p = \rho g h \qquad (1)$$

where: p is the pressure
ρ is the density
g is the acceleration of gravity
h is the column height

This equation is known as the *hydrostatic equation*. Using average figures in the cgs system, the pressure at sea level will be around one million dynes per square centimeter, which has been defined in engineering circles as a *bar*. Since the variation of pressure is quite small compared to this large unit, the renowned meteorologist Bjerknes suggested the use of the millibar, or one one-thousandth of a bar. Now a normal pressure reading would be in the vicinity of 1 000.00 millibars (mbs). The standard atmosphere referred to before uses the following values as standard for sea level: 1 013.2 mbs, 29.92 inches of mercury, or 760 millimeters (mm) of mercury. Notice that one is a true pressure measurement while the other two only refer to the height of the mercury column on a standard day.

There are two basic types of barometers, the *mercurial* barometer and the *aneroid* barometer, which are in popular use today for meteorological observation. The mercurial is slightly refined from Torricelli's initial effort, but still is an open cistern, subjected to ambient air pressure, supporting the column of mercury. In the case of a shipboard installation,* the barometer is suspended in gimbals, or a two-way hinge that will permit it to stabilize itself with relation to a vessel in motion. A thermometer is attached to enable an observer to take simultaneous temperature readings. This is to facilitate the application of an expansion-contraction factor to both the mercury and the glass tube in which it is housed. Normally an instrument correction is also applied, which is peculiar to each particular instrument and has been precomputed by the manufacturer. Because the earth is an oblate spheroid instead of a true sphere and the distance to the center of mass is not uniform over the surface, a compensation must also be made for this variation in gravity. Finally to make all readings comparable, regardless of the elevation in the atmosphere (that is the elevation of the observing station), all readings must be reduced to a standard reference, sea level. After these corrections have been applied, the pressure reading is useful for meteorological purposes.

A more simple instrument is the aneroid barometer, consisting of a partially evacuated cylinder with accordian-type sides. A spring is mounted inside to keep the evacuated container from collapsing. In response to changes of atmospheric pressure, the dimensions of this container will change. By a simple set of cranks and levers, these changes are converted to needle movements across the face of a dial. A bimetallic arrangement is built into the device so that temperature compensations are automatically made. Since the effect of gravity is not important for this instrument, the only corrections that must be made are the instrument correction and the allowance for the observer's elevation above sea level. It would be a simple process to graduate the scale of an aneroid barometer in thousands of feet, instead of millibars, and this is exactly what has been done. When installed in an airplane, it is called an altimeter. The pilot must make allowances for the temperature change to correct this to pressure altitude, but this correction is small.

By installing a pen on the indicating needle of an aneroid barometer

* Most Navy ships do not carry mercurial barometers due to the relative ease of use and maintenance of the aneroid type.

Figure 4-5 Aneroid Barometer

and letting it trace a line on a slowly rotating drum, a time record of the pressure will be obtained. This is called a barograph. A very important consideration in weather forecasting is the change or tendency of the pressure. The barograph gives a graphic presentation of this necessary parameter.

409 Summary. In summary this chapter has tried to outline the many divisions of the atmosphere, both near and far. Some important spheres are as follows:

ACCORDING TO TEMPERATURE
- Troposphere
- Stratosphere
- Mesosphere
- Thermosphere

ACCORDING TO COMPOSITION
- Homosphere
- Heterosphere (Exosphere)

ACCORDING TO SPECIAL CHARACTERISTICS
- Ionosphere
- Magnetosphere
- Chemosphere (Ozonosphere)

Finally, consider the pressure exerted by the tremendous mass of the air above. The hydrostatic equation and the universal gas law control its behavior. It decreases rapidly with altitude, exhibiting half of its mass in the first 18,000 feet, yet showing a measurable pressure of 10 millibars as high as 100,000 feet. The atmosphere has, for aeons, beckoned to man, and man has answered the call. From the time that Greek storytellers invented the tale of Daedalus and his son Icarus, man has been fascinated by the thought of investigating the upper reaches of the atmosphere in person. (These two intrepid aviators constructed wings of feathers and wax to escape the island of Crete. Alas, Icarus flew too near the sun, which melted the wax wings and he plunged to his death.) Man has tamed the upper air and is now in the process of invading the vast reaches of the boundless space, yet he can go only by taking a part of the atmosphere along with him, to breathe.

Additional Reading

Baker, B. B., Jr., Deebel, W. R. and Geisenderfer, R. D., *Glossary of Oceanographic Terms*, 2nd Ed. U. S. Naval Oceanographic Office, Govt. Printing Office, Washington, D. C., 1966.

Barry, R. G., and Chorley, R. J., *Atmosphere, Weather, and Climate*, Holt, Rinehart, and Winston, 1970.

Byers, Horace R., *General Meteorology*, 3rd ed., McGraw-Hill, Inc., 1959.

Glasstone, Samuel, *Sourcebook on the Space Sciences*, D. Van Nostrand Co., Inc., Princeton, N. J., 1965.

Haltiner, George J., and Martin, Frank L., *Dynamical and Physical Meteorology*, McGraw-Hill, Inc., 1957.

Huschke, Ralph, E., *Glossary of Meteorology*, American Meteorological Society, Boston, Mass., 1959.

Landsberg, H. E., "The Origin of the Atmosphere," *Scientific American*, August, 1953.

Petterson, Svere, *Introduction to Meteorology*, McGraw-Hill, Inc., 1958.

Riehl, Herbert, *Introduction to the Atmosphere*, McGraw-Hill, 1965.

White, Stephen R., "The Earth's Radiation Belts," *Physics Today*, October, 1966.

———, *The Upper Atmosphere*, NWRF 26-0665-106, Naval Weather Research Facility, June, 1965.

CHAPTER FIVE

Sun: The Energy Source

501 Earth, Water, Air, and Fire. During the latter half of the fifth century B.C. a Greek philosopher named Empedocles of Acragas postulated the concept of four primordial elements. According to Empedocles these "roots of all things" were earth, water, air, and fire. If we would accept a broad interpretation, it would appear that he was the first serious student of the air-ocean environment. Indeed, our subject matter is the same as his since the preceding three chapters have been concerned with earth, water, and air, and this one will discuss "fire." Twenty-five centuries of science notwithstanding, this entire book has been, and will be, involved with the examination of earth, water, air, and energy and the interaction of each with the other. So that the crude (by present standards) simplification of Empedocles turns out to be a very valuable description of the marine environment.

Courtesy: Robert deGast

502 Energy and Its Transport. Energy, per se, is of little interest or importance. When energy is *transported*, people think; trees grow; and storms develop. Consequently the various means by which energy is moved from one place to another will be examined in some detail.

There are many kinds of energy, from that contained in breakfast cereals to that contained within the confines of a full-fledged hurricane. However, with only one possible exception (nuclear energy), all energy found within the confines of the oceanic-atmospheric system is in the form of heat or has resulted from a heat exchange.

The three methods of heat transfer commonly found in nature are *conduction*, *convection*, and *radiation;* the relative importance of each depends on the type of physical process involved.

If the heat is transported by means of molecular agitation without any motion of the heat-containing material, this is called *conduction.* In terms of total amount of heat energy exchanged in nature, conduction is the least important of the three processes mentioned and can usually be neglected when compared with the other two.

Consider now a process whereby a heat-containing fluid moves and in so doing carries heat along with it. This is how most of the water in a pan placed on a stove is heated. The heat travels from the flame through the pan by conduction to the water right on the bottom. This water is heated, becomes less dense and rises, warming the upper water as it moves. A convection current is set up. When heat is carried by a fluid in motion, this process of heat transfer is called *convection* if the motion is vertical. If the heat is transferred by horizontal motion, the process is termed *advection.*

503 Radiation. In convection some sort of a material substance was required to effect the transfer of heat. If heat is transported by the process of radiation, no material is necessary since heat is a member of

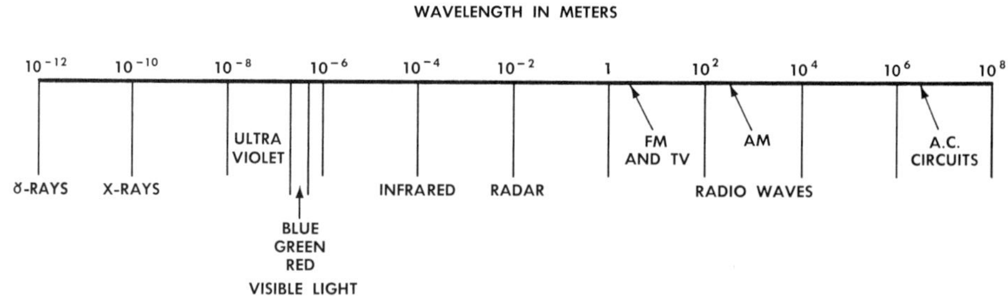

Figure 5-1 The Electromagnetic Spectrum

the wave-energy family called the *electromagnetic spectrum*. This family, including such diverse forms of energy as x-rays, light waves, and radio waves, all have the property of being able to travel through a vacuum with the speed of light (3×10^8 meters/sec or 186,000 miles/sec). One of the major differences between the various portions of the electromagnetic spectrum is that of wavelength. Radio waves, for example, have wavelengths between about 1 meter (39.37 inches) and 20 kilometers (12.5 miles) while infrared, commonly called heat, has wavelengths between 10^{-6} meters (.00004 inch) and 1 millimeter (.04 inch). The wavelength of visible light varies between about $.35 \times 10^{-6}$ meters (violet light) and $.75 \times 10^{-6}$ meters (red light).

Since heat may be considered to be an electromagnetic wave, the transmission of heat by means of these waves is an acceptable process, and one which is very important in nature. If heat is exchanged by means of electromagnetic waves, this process is called *radiation*.

The process of radiation is important in the natural exchange of heat. All bodies having a temperature greater than 0°K (−273°C) emit some radiation; its amount and character is determined by the absolute temperature and the material making up the body. This phenomenon is utilized in infrared photography and in temperature determination by infrared techniques. For example, airborne radiation thermometers (ART) are being utilized to locate the changing positions of the Gulf Stream system on a routine basis.

The relationship which most closely describes the character of radiation for a radiating body was suggested by Max Planck in 1900. Planck's Radiation Law is given by

$$\frac{E}{\Delta\lambda} = \frac{\lambda^{-5} C_1}{(e^{C_2/\lambda T} - 1)} \tag{1}$$

where E = The energy emitted per unit area per unit time
$\Delta\lambda$ = A unit wavelength band
λ = Radiation wavelength
C_1 and C_2 are constants
T = Absolute temperature.

It may be seen that the energy emitted per unit time per unit area within a given wavelength band is related to both the absolute temperature and the wavelength. A few plots of Planck's Law are shown in Figure 5-2 wherein $E/\Delta\lambda$ is plotted against λ. Note that, as the temperature is increased, two effects may be seen:

1) the wavelength of maximum emission decreases, and
2) the total amount of energy (the area under the curve) increases.

These two effects may be analyzed in a more quantitative manner by

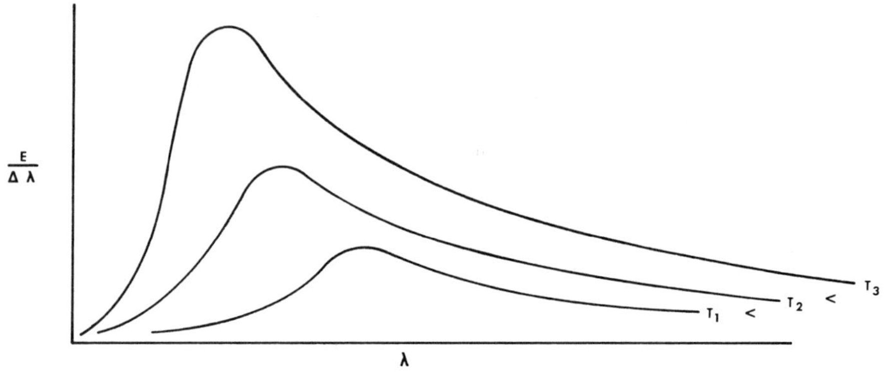

Figure 5-2 Planck's Radiation Law

operating on Planck's equation. Since the wavelength of maximum emission is the maximum point of the curve, all that is needed to determine the variation of this wavelength with temperature is to take the first derivative of Planck's Equation $d(E/\Delta\lambda)/d\lambda$, set it equal to zero, and solve for the value of λ at the point of maximum energy emission. When this is done, the resulting expression is called Wien's displacement law:

$$\lambda_{max} = \frac{2897}{T} \tag{2}$$

where λ_{max} is given in micrometers (10^{-6} meters) and T in degrees Kelvin.

On the other hand, if a relationship for the total amount of energy emitted by a body per unit area per unit time is desired, Planck's equation may be integrated:

$$\int_0^\infty \left(\frac{E}{\Delta\lambda}\right) d\lambda = E \tag{3}$$

When this is done, the result is identical to that developed independently by Stefan and Boltzmann and is consequently known as the Stefan-Boltzmann Law:

$$E = \sigma T^4 \tag{4}$$

where σ is simply another constant called the Stefan-Boltzmann constant.

In concluding this section on the radiation laws, a few more parameters are defined which will be of interest as our discussion proceeds. It is convenient to first define an object called a *black body*. If a body radiates according to Planck's Law at each and every wavelength, it is called a black body. The term *gray body* is sometimes applied to materials which emit an amount somewhat less than predicted by Planck's Law.

The ratio of the radiation actually emitted by a body at some wavelength to that of a black body at the wavelength is called the *emissivity* (e). Thus the emissivity may vary from 0 to 1 in value, and the e of a black body is fixed at unity.

Similarly, the ratio of the radiation actually absorbed by a body to that which is incident is called the *absorptivity* (a). Consequently the absorptivity may vary from 0 to 1 in value while (a) of a black body is also equal to unity.

In general, when a body is a good absorber of radiation it is also a good emitter. This is expressed as Kirchhoff's Law:

$$e = a \tag{5}$$

When radiant energy is incident upon an object some of the radiation is absorbed by the body and some is reflected. The ratio of the amount of radiant energy reflected to that incident is called the *reflectivity*. In the environmental sciences, reflectivity is usually called *albedo* (A), where a perfect reflector has an albedo of unity.

At this point it should be obvious that the sum of the absorptivity and the albedo must always be equal to unity for any opaque body. This is true since all energy must be taken into account and the only physical processes affecting the incident radiation are absorption and reflection. Thus:

$$a + A = 1. \tag{6}$$

504 Heat Budget. About 93,000,000 miles from the earth is the center of our solar system, the sun. From the sun radiates all the energy necessary to drive wind and current systems and, ultimately, to sustain life. The source of this energy is apparently some sort of a thermonuclear reaction within the sun which produces a solar surface temperature of about 6,000°Kelvin. From Wien's displacement law (equation 2), we may easily calculate the wavelength of maximum emission of the sun, assuming it to be a black body:

$$(\lambda_{\text{max}})_{\text{sun}} = \frac{2{,}897}{6{,}000} = 0.482 \text{ micrometers.}$$

This peak solar emission corresponds to a blue-green color in the visible spectrum.

A similar calculation for the earth-atmosphere system may be made assuming it to have an average temperature of about 300° Kelvin.

$$(\lambda_{\text{max}})_{\text{earth}} = \frac{2{,}897}{300} = 9.64 \text{ micrometers}$$

This peak terrestrial emission is located quite far beyond the visible spectrum in the infrared. Due to the fact that the wavelength of maximum emission of solar radiation is so much smaller than that for terrestrial radiation, solar radiation is often called *short-wave radiation*, while that emitted by cooler bodies such as the earth-atmosphere system is termed *long-wave radiation.* These two emission curves are shown in Figure 5-3; note the logarithmic scale for $E/\Delta\lambda$. Due to the difference in solar and terrestrial temperatures there is a tremendous disparity in the amounts of energy emitted. A quantitative comparison may be made by utilizing equation (4):

$$\frac{E_{\text{sun}}}{E_{\text{earth}}} = \frac{T_{\text{sun}}^4}{T_{\text{earth}}^4} = \frac{(6{,}000)^4}{(300)^4} = 160{,}000.$$

In other words, the energy emitted *per unit area* per unit time by the sun is 160,000 times as great as that radiated by the earth. And since the total surface area of the sun is 10,000 times greater than that of the earth, the *total* energy emitted by the sun is 1.6 billion times as great as that emitted by the earth.

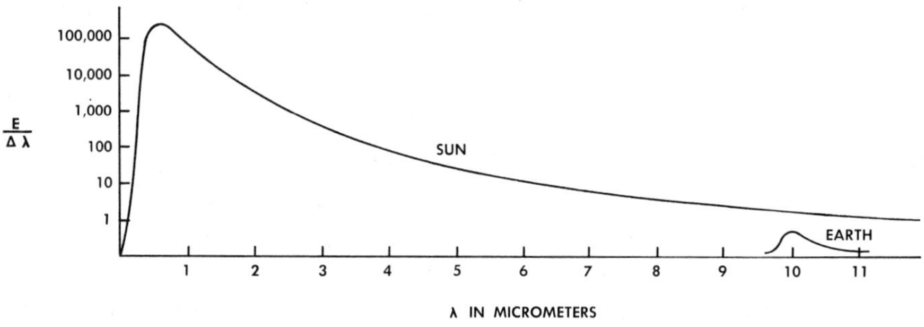

Figure 5-3 Relative Radiation of the Sun and Earth

However, since the earth only receives such a small portion of the sun's total radiation (Fig. 5–4), it turns out that *the amount of solar energy absorbed by the earth-atmosphere system is exactly equal to that emitted by the earth-atmosphere system.* This must be true if the average temperature of the terrestrial environment is to remain constant from year to year. If the energy input were greater than that reradiated back into space, average atmospheric temperatures would of necessity increase; and if more energy were radiated into space than was received, average temperatures would decrease. Although there is some evidence that average temperatures have been increasing slightly over the past

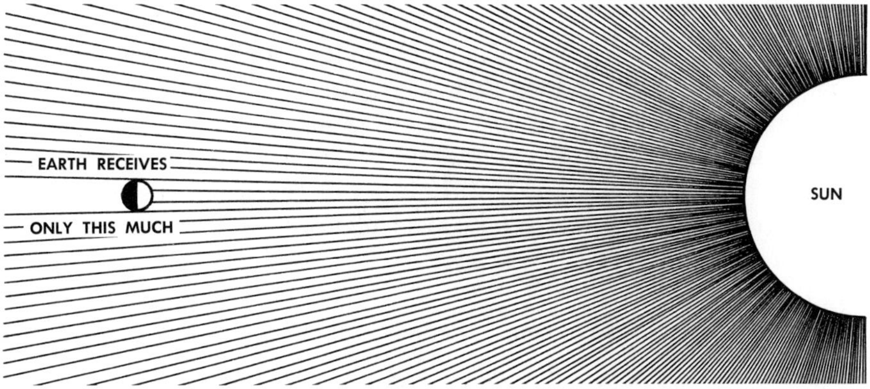

Figure 5-4 Solar Radiation Received by the Earth

fifty years or so, the change is so small, and the time span is so short, relatively speaking, that one may still assume the concept of heat balance to be valid.

505 Losses. As electromagnetic energy from the sun enters the earth's atmosphere, certain losses start to accrue since the processes of absorption, scattering,* and reflection tend to decrease the amount of energy available at the earth's surface. To illustrate, assume that 100 units of solar radiation reach the top of the atmosphere. Due to re-

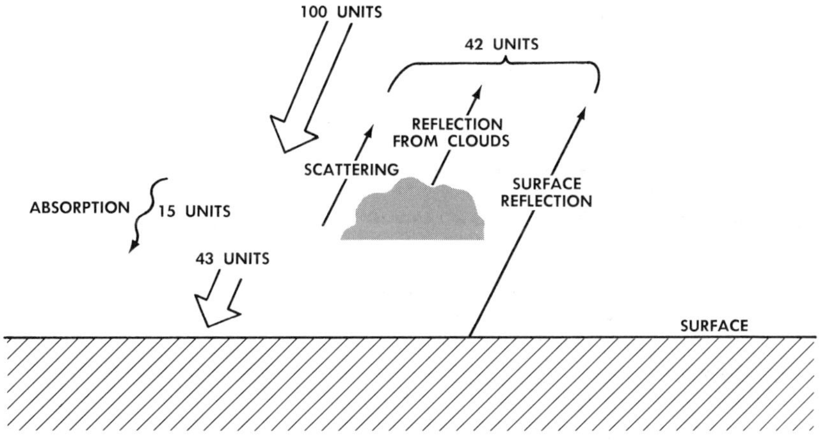

Figure 5-5 Atmospheric Radiation Balance

* See Chapter 8 for a more complete discussion of absorption and scattering.

flection from clouds, reflection from land and water at the earth's surface, and scattering by particulate matter such as dust particles, along with the air molecules themselves, about 42 units of radiation are lost back into space.

Due to absorption by water vapor and other gases in the atmosphere, another 15 units are lost, so that only 43 of the original 100 units of short-wave radiation remain to be absorbed at the earth's surface.

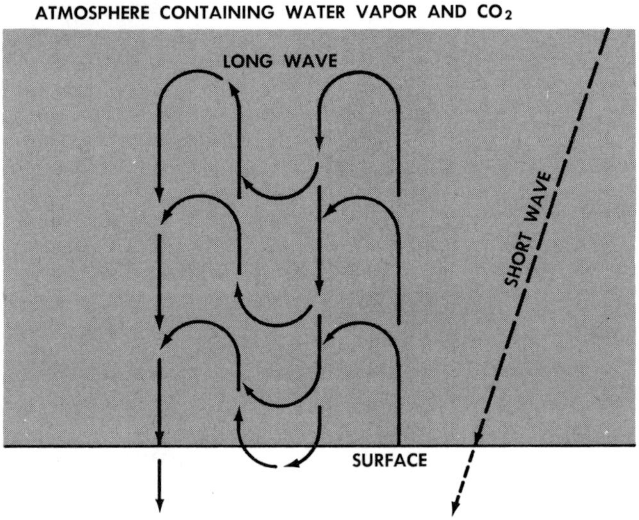

Figure 5-6 Greenhouse Effect

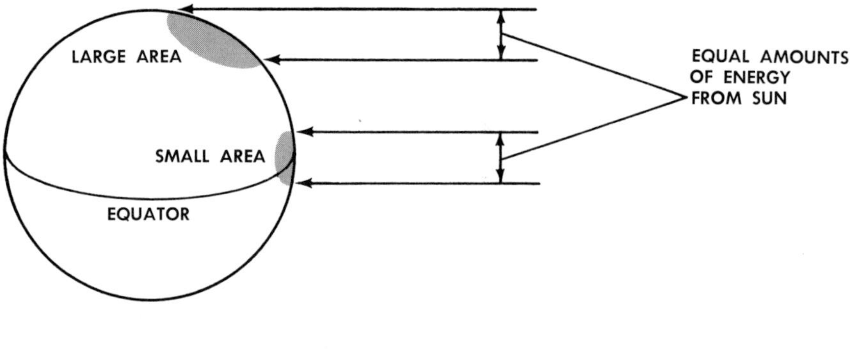

Figure 5-7 Uneven Heating at Different Latitudes

The losses due to scattering are selective in that the shorter wavelengths are more affected than the longer ones. This has the effect of scattering blue light 10 to 15 times as much as red light. Most of the absorption within the atmosphere is also selective in nature. Ozone, for example, absorbs very strongly in the ultraviolet region of the spectrum, while water vapor and carbon dioxide absorb very strongly in the infrared.

Water in the liquid state also absorbs very strongly in the infrared. This, coupled with the fact that about 40 percent of the solar radiation reaching the earth's surface is in the infrared, results in more than half of the solar energy input being absorbed in the upper few meters of the world ocean. For this reason about 55 percent of the solar energy entering the hydrosphere is immediately utilized for evaporation, leaving less than half to maintain the water temperature. High infrared absorption by water and the large resulting evaporation are in large measure responsible for our weather.

Whenever an absorption of energy takes place, except for the latent heat associated with change of state, there is always an associated increase in temperature. But, whenever the temperature is raised there is always an associated increase in reradiation. In other words, a body which absorbs energy will tend to reradiate it until an equilibrium condition is reached where the amount of energy absorbed is equal to that emitted. Thus the earth is receiving energy continuously from the sun and, in turn, reradiating energy at some rate determined by the amount of solar radiation absorbed. Remember, however, that the energy received from the sun is short-wave and that which is re-emitted is long-wave due to the temperature difference between earth and sun.

As was previously seen, about 15 of the 100 units of short-wave energy reaching the atmosphere were lost due to absorption. However, the atmosphere is much more opaque to long-wave radiation, mainly because of its water vapor, carbon dioxide, and to some extent, ozone content. This state of affairs results in the *greenhouse effect*. Short-wave (solar) radiation is easily passed through the atmosphere, while long-wave (terrestrial) radiation is absorbed by the atmosphere and reradiated back down towards earth. This is especially noticeable when there is a cloud cover at night, tending to keep temperatures somewhat higher than they would be with a clear sky.

506 Uneven Heating. Up to this point the earth-atmosphere system has been treated as a unit, describing incoming and outgoing radiation in terms of the averages. This is a valid approach in certain areas; but if one is to be concerned with weather and the dynamic ocean, it is necessary to leave the realm of averages and examine individual conditions. Personal experience shows that the reaction of the earth to solar radiation varies greatly from spot to spot. Swimming is possible during December in Miami, but never in the Arctic; people ski during July in South America, but never in Pago Pago. It is fairly obvious that the surface of the earth is *unevenly* heated by solar radiation.

This uneven heating comes about from three basic causes. The first of these is the *shape of the earth*. Since the earth is a sphere, solar radiation strikes the earth at an ever increasingly oblique angle as one moves away from the equator. As may be seen from Figure 5-7 the same amount of energy is available at higher latitudes, *but it is spread out over a greater area*, so that the energy striking a unit surface area decreases with increasing latitude. At higher latitudes, then, the sun is lower in the sky.

The second cause of uneven heating is the *inclination of the earth's axis* of rotation. Since the axis is always pointed approximately toward the same spot in space, the relative position of the sun with respect to any selected latitude will vary over the course of a year. Note in Figure 5-8 that during the northern hemisphere summer, the earth is actually somewhat farther from the sun than during the winter. The increase in solar radiation felt during the summer season is produced by the direction of the sun's rays approaching the vertical, not the change in earth-sun distance.

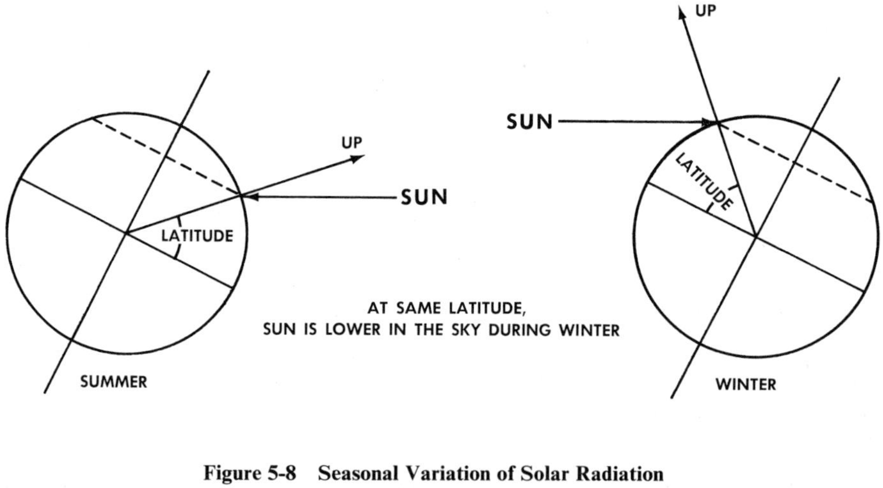

Figure 5-8 Seasonal Variation of Solar Radiation

Lastly, uneven heating may be produced by the proximity of two areas having *differing thermal properties* or albedos. Take, for example, a coastal area wherein a land mass having a relatively low heat capacity is adjacent to a water mass having a relatively high heat capacity. Because of this difference in heat capacity, the same amount of solar energy impinging on both will cause the ground to reach a higher temperature than the water. Conversely, at night, the ground will decrease its temperature more rapidly than the water as both emit long-wave radiation.

It should be noted that mobility is a major factor in the ability of the ocean to store greater amounts of heat than land. The transfer of heat through a solid material such as rock is limited to conduction; however, a fluid medium such as water may also transfer heat by convective processes. Because of this, large bodies of water usually have a moderating effect on the climate of the surrounding land mass, tending to keep the temperature lower during the summer and higher during the winter.

In addition to differences in heat capacity, uneven heating may also be produced by differences in albedo. Undoubtedly one of the major reasons the ice caps do not melt during the summer months is the extremely high albedo of snow—about .75 to .80. This must be the case since during the summer months the total amount of solar radiation received during a 24-hour period is greater at the North Pole than it is at the Equator. Of course, there is no night during this period at higher latitudes which accounts for the greater total radiation received. The principle of reflectivity is utilized in ice-breaking operations where ice may be initially weakened by dumping oil (dirty lubricating oil, preferably) on its surface. This artificially produces an albedo difference, raising the temperature of the ice beneath the oil due to the higher absorptivity of this region, thus tending to melt it.

507 Summary. The source of all energy is the sun. Just how solar energy is utilized to drive the hydrosphere-atmosphere machine is something else again. But two mechanisms of apparent importance discussed in this chapter are the large amount of solar heat utilized for evaporation and the various methods of producing uneven heating. The wind blows, the sea moves, the atmosphere fairly splits asunder, unable to contain the energy of a common thunderstorm—all these and more are attributable to uneven heating and evaporation.

Additional Reading

Blair, T. A. and Fite, R. C., *Weather Elements*, 5th ed., Prentice-Hall, Inc., 1965.

Byers, H. R., *General Meteorology*, 3rd ed., McGraw-Hill, Inc., 1959.

Defant, A., *Physical Oceanography*, Vol. I, Pergamon Press, 1961.

Dietrich, Gunter, *General Oceanography*, John Wiley & Sons, 1963.

Donn, W. L., *Meteorology*, 3rd ed., McGraw-Hill, Inc., 1965.

Neumann, G. and Pierson, W. J., Jr., *Principles of Physical Oceanography*, Prentice-Hall, Inc., 1966.

Öpik, E. J., "Climate and the Changing Sun," *Scientific American*, June, 1958.

Petterson, S., *Introduction to Meteorology*, McGraw-Hill, Inc., 1958.

Plass, G. N. "Carbon Dioxide and Climate," *Scientific American*, July, 1959.

Riehl, H., *Introduction to the Atmosphere*, McGraw-Hill, Inc., 1965.

Sverdrup, H. V., Johnson, M. W., and Fleming, R. H., *The Oceans*, Prentice Hall, Inc., 1942.

Von Arx, W., *Introduction to Physical Oceanography*, Addison-Wesley Publishing Co., Inc., 1962.

CHAPTER SIX

Temperature Structure of the Ocean

601 Surface Temperature. The last chapter concerned that portion of solar energy which manages to find its way through the atmosphere, is absorbed within the ocean, and utilized to maintain the oceanic temperature. Variation of temperature in the ocean is a function of both position and time; there is more heat input at lower latitudes than at higher latitudes and more heat input during the summer than during the winter. Therefore, one would expect oceanic temperatures to be greater near the equator than they are near the poles. Also in mid-latitudes, seasonal variation would be expected to produce warmer temperatures in summertime than in wintertime. In general, these expectations are realized.

A long-term record of the surface oceanic temperature shows its variation in both space and time. Referring to Figure 6-1, which is a plot of yearly average surface temperatures, a number of features are immediately obvious. In the first place, as predicted, surface waters are indeed warmer near the equator than they are near the poles. How-

Bathythermograph breaks the surface

Figure 6-1 Yearly Average Surface Temperature of the Oceans (°C)

ever, an anomaly may be noted since the maximum temperature is not found occurring at the equator, even though there is a greater average radiation input per unit time there than anyplace else on earth. The average position of the maximum oceanic surface temperature, called the *oceanographic thermal equator*, is located at about five to ten degrees north latitude.

Although not exactly coinciding with the oceanographic thermal equator, the location of the highest surface air temperature for a given period of time (*the meteorological thermal equator*) is also usually somewhat north of the equator. This is one result of the Northern Hemisphere being somewhat warmer than the Southern Hemisphere.

One might ask why the Northern Hemisphere is warmer, and two fairly reasonable answers may be given to this question. The first reason is the fact that there is more water in the Southern Hemisphere than the Northern Hemisphere. As a matter of fact, in the Northern Hemisphere about 39 percent of the total surface area is covered by land and about 61 percent is water; whereas in the Southern Hemisphere only about 19 percent of the total area is land, and 81 percent is oceanic in nature. This large amount of water coupled with the aforementioned high specific heat capacity of water allows the Southern Hemisphere to absorb great amounts of heat without correspondingly large temperature changes. In other words, the same amount of heat in the form of solar radiation striking the Southern Hemisphere will cause less of a temperature increase in the southern oceans than in the northern.

Another reason for the displaced oceanographic thermal equator may perhaps be explained by the presence of the continent of Antarctica. Contrary to popular belief (due probably to the type of projection used in most maps), Antarctica is a rather large continent. It is approximately equal in size to all of North America with the exception of Mexico and Alaska. This large land mass is continually covered with a very thick sheet of ice. So that in addition to the greater amount of water absorbing heat without changing the temperature, the earth has this large icebox in the Southern Hemisphere also tending to keep the temperature down. In any event, the Northern Hemisphere is warmer than the Southern Hemisphere and the mean position of the oceanographic thermal equator is displaced somewhat north of the geographic equator.

One interesting aspect of this whole problem of the conversion of heat energy and temperature variation is the relative effect of heat transferred by the process of conduction. It was indicated in the previous chapter that conduction was relatively minor in importance as far as heat transfer is concerned. Some idea as to the order of magnitude involved may be obtained by looking at a numerical example. If a hypothetical ocean has a temperature of 0°C at all layers except the surface, and the surface of that ocean is maintained at 30°C by the radiant heat of the sun, the amount of heat required to keep the surface at this elevated temperature is very small because of the slow rate at which heat is conducted by water. The rate of conduction is so slow, in fact, that after a period of 1,000 years the temperature at a depth of 300 meters would be raised only from 0°C to 3°C, if there were no convection. Obviously this is not the case. By virtue of convection currents set up within the medium, large amounts of heat are carried to much of the oceanic volume.

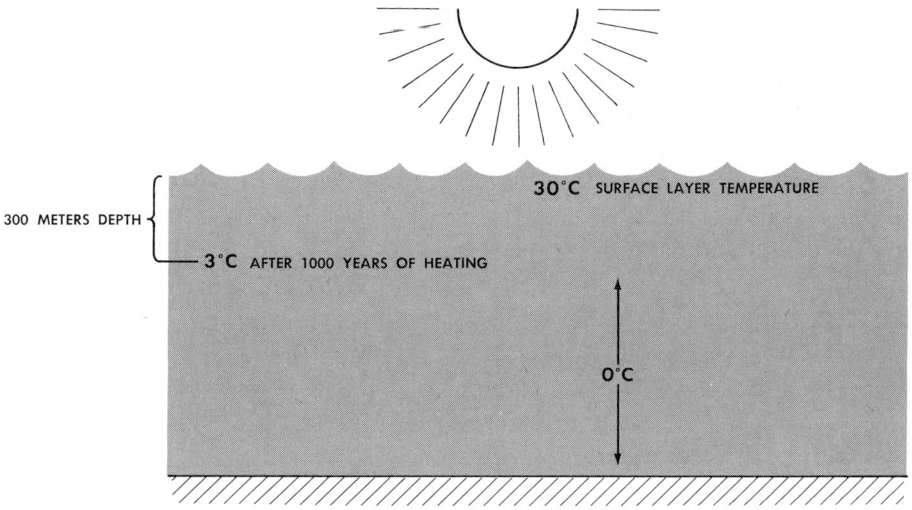

Figure 6-2 The Slow Rate of Heat Transfer by Conduction

Before looking at the vertical variation of temperature, examine the effect of surface currents on surface temperature variation. If there were no surface currents, the isotherms would be parallel to lines of latitude. That is, the temperature of the surface waters would be the same at equal latitudes throughout the world. However, surface currents do exist, and the surface isotherms are often not parallel to lines of latitude. Referring to Figure 6-1 again, observe two deviations from the ideal pattern. One is the fact that the isotherms are often kinked along oceanic edges. The western side of the North Atlantic, for example, exhibits kinking such that warmer waters are found farther north than one might expect. On the eastern side of the North Pacific the isotherms are kinked so that colder water is found farther south than one would normally expect. These displacements are a result of the permanent surface current systems which will be examined in detail later on.

In addition, the surface currents tend to bunch the isotherms at higher latitudes, so that the rate of change in surface temperature increases with latitude the closer one gets to the poles. For example, within the latitude range of from 0° to 20° North and South there appears to be an average temperature change of about 2° to 3°C. However, within the latitude range from 20° to 40° North and South there appears to be a temperature change of between 5°C and 10°C, and from 40° to 60° North and South latitude there is an average temperature change of about 20°Celsius. It would appear that the farther away from the equator one gets, the more rapidly the surface temperature changes.

Temperature more uniform at equator with latitude

602 Temperature Differences. Another thermal aspect of changing latitude is involved with the annual range in temperatures to be expected as a function of latitude. One would expect that the range in temperatures of the surface waters would be small at both the equator and very high latitudes, since it is always warm at the equator and it is always cold in polar regions. And this is about the way it turns out. Referring to Figure 6-3, it is seen that the ranges are smaller at higher latitudes and greatest in mid-latitudes where seasonal effects are felt, and that the minimum range occurs at the oceanographic thermal equator.

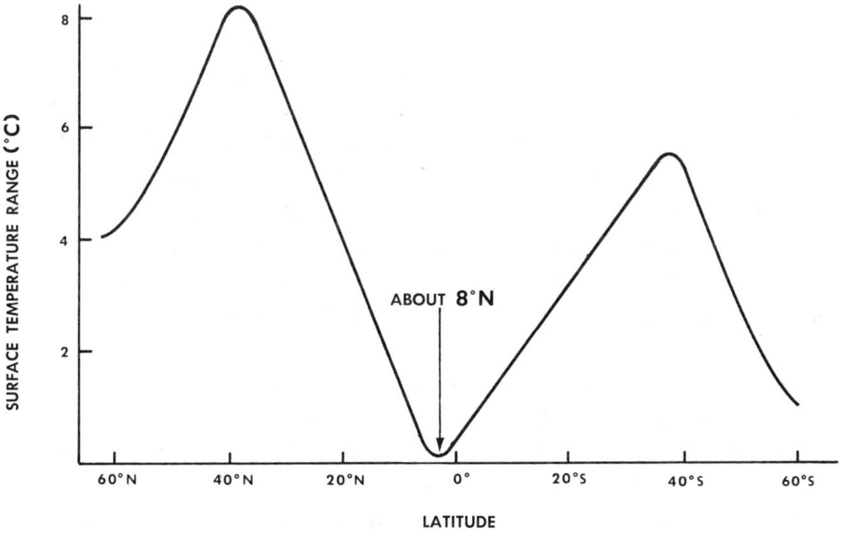

Figure 6-3 Yearly Range in Surface Temperature in the Atlantic Ocean

Similar variations occur when one discusses the *diurnal* range in temperature. It is not surprising that smaller changes occur in equatorial and polar regions with maximum change experienced in the middle latitudes. This daily variation is similar to that of the annual variation, only the expected ranges vary from about 0.1°C to about 2°C.

Another effect of the difference in mobility and specific heat capacity of water and land shows up in air/sea and air/land temperature differences. In Table 6-1 average air temperatures over land (along the 120°E meridian) and water (along the 20°W meridian) between 20°N and 80°N are tabulated. It may be seen from Table 6-1 that in the winter the average air temperature over land is about −16°C whereas over water the average air temperature is about 6.3°Celsius. Notice that the air over land is colder than the atmosphere over water in the winter. In the summer sea air is cooler than land air. The really big difference shows up when one compares the variation in air temperature over the land for an entire year (35°C) whereas over the ocean it is only about 8°Celsius. In summary then, the oceans tend to moderate atmospheric temperatures. The general effect is to produce more moderate climates in coastal areas as compared to those farther inland.

Table 6-1 Average Air/Sea and Air/Land Temperature Differences for Mid-Latitudes

	AIR TEMPERATURE OVER LAND	AIR TEMPERATURE OVER SEA	DIFFERENCE
Winter	−15.9°C	6.3°C	22.2°C
Summer	19.4°C	14.6°C	− 4.8°C
Winter——Summer Temperature Seasonal Difference	35.3°C	8.3°C	

603 Temperature Variation with Depth. Leaving the surface to examine the variation of temperature in the third dimension, i.e., depth, one finds that, to a very good approximation, the ocean may be thought of in terms of two basic layers. In the surface region there is a layer of relatively warm water called the *mixed layer*, and down near the bottom there is a layer of relatively cold water called the *deep layer*. In between these two layers there is a transition area in which the temperature changes rapidly with depth called the *main thermocline.* Figure 6-4 shows this layer-cake structure. Note that in the higher latitudes, the deep layer comes all the way to the surface so that here one would expect the temperature to be consistently cold from top to bottom.

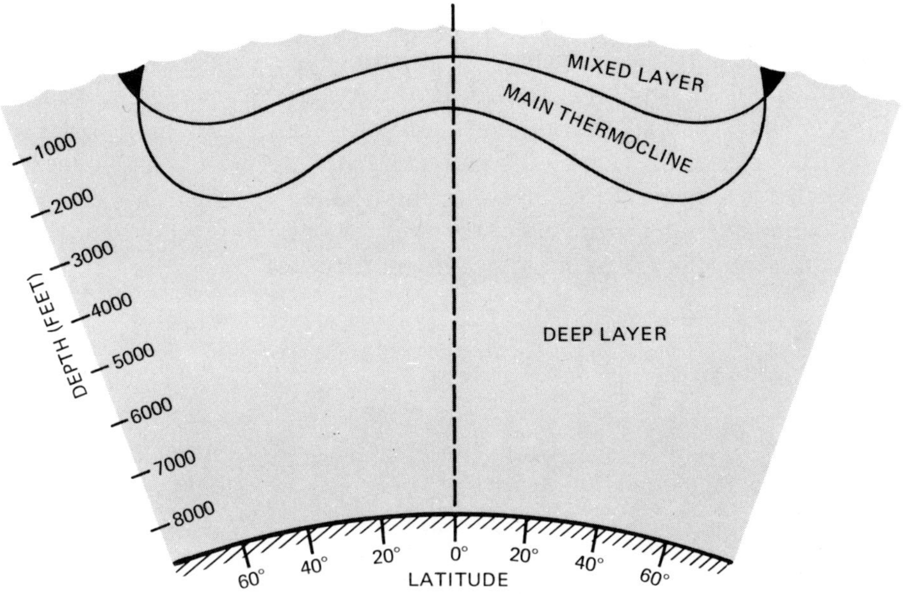

Figure 6-4 The Three-Layered Ocean

It should be emphasized that the mixed layer has a relatively constant temperature for any particular geographic location desired. However, the temperature of the mixed layer at 30°N as compared to that at the equator is somewhat different. As was previously seen, the surface temperature of the oceans varies with latitude, so that the mixed layer is not a homogeneous mass of water all having the same temperature. Nevertheless, at any particular geographic location the vertical change in temperature is quite small within the mixed layer.

If the variation of temperature with depth is plotted for three typical latitudes: high latitude (55°–90°), mid-latitude (30°–55°) and low latitude (0°–30°),* the three curves in Figure 6-5 result. At low latitudes the surface temperature is warmer than it is at mid-latitudes and also both the mixed layer and the main thermocline are thinner than they are at mid-latitudes. At high latitudes there is no change in temperature with depth.

In addition to this vertical variation of temperature with depth there is also a temporal or a time variation superimposed on the structure

* Latitude boundaries are somewhat variable; these are only representative figures.

given above. The plot in Figure 6-5 of the temperature variation with depth for mid-latitudes is actually that for the winter season. Examining the thermal structure for some other season one finds that the mixed layer changes its temperature structure only in mid-latitudes. This change in character of the mixed layer in mid-latitudes is shown in Figure 6-6. Temperature is constant with depth during the wintertime; but with the approach of spring, the surface layers become warmed and an additional thermocline called the *seasonal thermocline* starts to appear. This seasonal thermocline becomes deeper and more marked as summer passes; in the fall, it becomes slightly deeper but less marked; and finally as winter returns, it disappears. As may be seen, this seasonal thermocline averages somewhere between 150 and 300 feet in depth and may represent as much as 5° to 10°C in temperature difference.

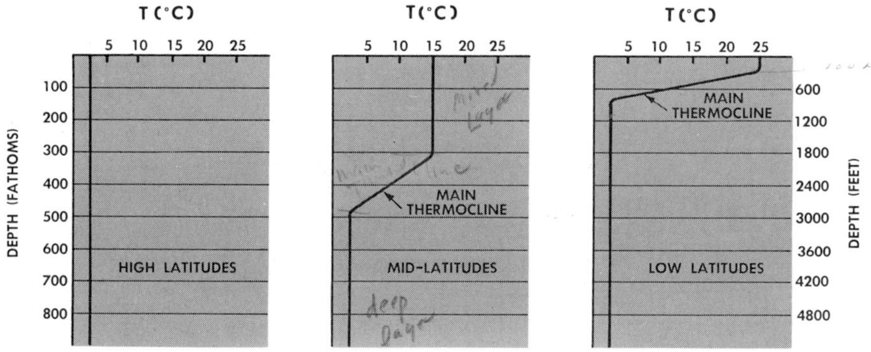

Figure 6-5 Idealized Winter Temperature Variation With Depth at Three Latitudes

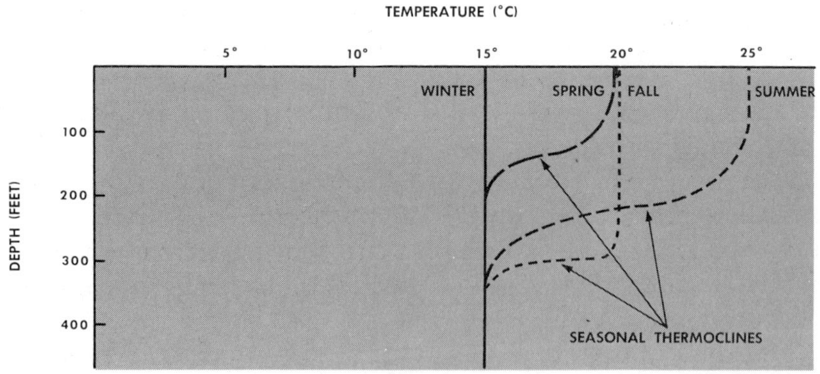

Figure 6-6 Typical Seasonal Variation of Surface Layer Temperature in Mid-Latitudes

Superimposed on this seasonal variation is also a diurnal or daily variation in temperature. Since the surface temperature varies during the day, being typically warmer in the late afternoon than in the early morning hours due to absorbed solar radiation, a fairly shallow additional thermocline develops. The *diurnal thermocline* is especially prominent during the spring, summer, and fall months. In Figure 6-7 progressive variations in vertical temperature structure are sketched for various times of the day. The absorption of solar radiation by the upper layers is readily apparent as the surface temperature increases to a late afternoon maximum and then decreases as the sun sets. Typical diurnal thermoclines may be as much as 30 feet deep and perhaps as much as 1° or 2°C in magnitude, though usually they are somewhat less. If a good breeze is blowing, the upper layers will become mixed, carrying the warm water to greater depths, having the effect of increasing the depth of the diurnal thermocline, but decreasing its magnitude, as the absorbed heat is spread out over a larger volume of seawater.

In summary, the deep ocean is characterized by a mixed layer, a main thermocline, and a deep layer. In mid-latitudes there may be and often are a total of three thermoclines, a *diurnal thermocline*, having a depth somewhere around 20 or 30 feet, a *seasonal thermocline* having a depth between 150 and 300 feet and a *main thermocline* at a depth between about 1,000 and 3,000 feet. As we shall see in Chapter 8, the path taken by sound energy is very strongly affected by the temperature variation. However, since most fish and ships operate in the upper regions of the ocean, the diurnal and seasonal thermoclines are of most interest.

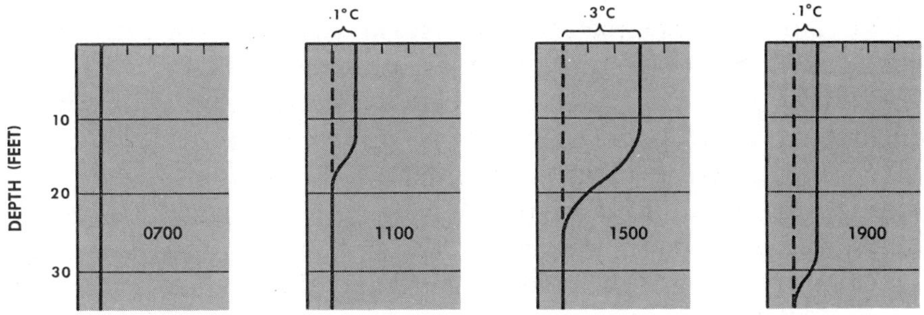

Figure 6-7 Typical Diurnal Warming Effects

604 Temperature Measurement. The measurement of surface water temperatures is a fairly simple procedure. Basically two methods are used. One is to simply utilize a ship's cooling water intake temperature since most larger vessels monitor this parameter. These temperatures are not extremely accurate but for many purposes they are satisfactory. More accurate surface temperature measurement may be made by the use of a bucket thermometer. Here a sample of water from the surface is captured, usually by means of a common bucket, and a simple thermometer is immersed in the sample.

Although the measurement of surface temperature is fairly straightforward, temperature measurement below the surface is somewhat more complicated. This may be accomplished in a number of different ways, the most popular of these being the use of the bathythermograph (BT).

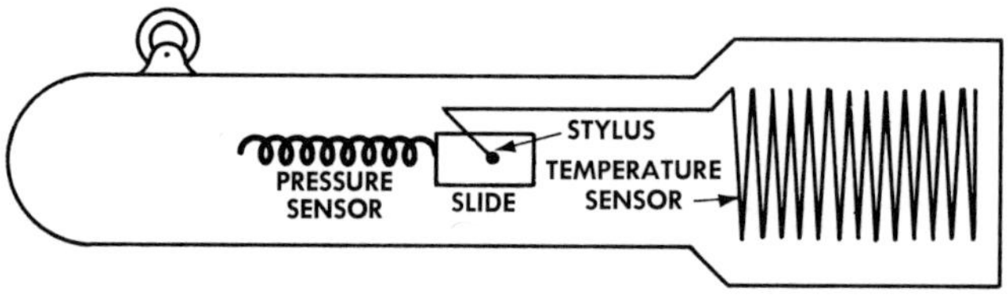

Figure 6-8 The Bathythermograph (BT)

A typical BT, as developed for the U.S. Navy, is diagrammed in Figure 6-8 and consists simply of a temperature-sensitive sensor which causes a stylus to move as the temperature changes and a pressure-sensitive sensor which causes a carriage to move at right angles to that caused by the temperature variation. On this carriage is placed a glass slide which is coated with a gold film, on which a stylus moves. As the device is lowered in the water, pressure (depth) and temperature variations cause motion of the slide carriage and stylus respectively, resulting in a plot of temperature versus depth. This slide may then be viewed against the background of a calibrated grid so that the actual temperature-versus-depth variation may be determined.

In this manner, the temperature structure may be obtained to about 800 feet in depth, with the major advantage being that the device may be used while the vessel is underway. The BT still has the disadvantage of requiring the ship to slow down to a speed of about 10 knots for optimum results. To meet this objection an expendable BT, shown in Figure 6-9, has been developed and is coming into greater use. This device is dropped over the side, and it telemeters back the temperature information along a very thin wire. The rate at which the unit sinks is known from previous calibration, so that the instrument depth may be determined as a function of time. At the end of its drop the wire is broken and the expendable BT sinks to the bottom. Both of these devices give a continuous plot of temperature versus depth.

For more accurate and deeper determinations, primarily for scientific use, other instruments are employed, usually in conjunction with water sampling gear. The basic device utilized by oceanographers to collect water samples is the Nansen bottle or some modification of it. A Nansen bottle is very simply a piece of pipe with a valve at each end, which may be lowered in the water to any desired depth with the valves open. When the bottle reaches the desired depth, a small brass cylinder (messenger) slides down the cable which trips the bottle and causes

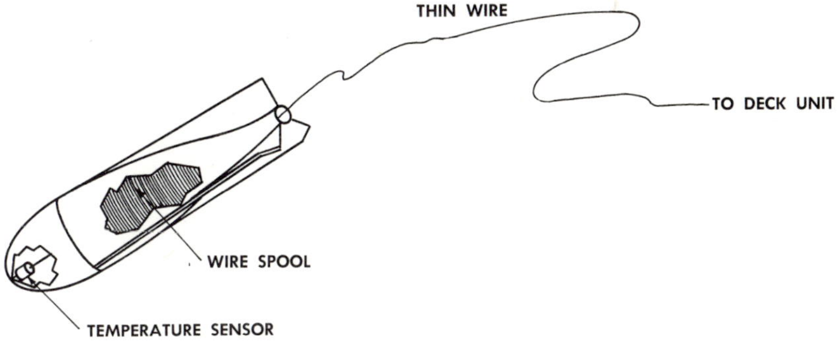

Figure 6-9 The Expendable Bathythermograph (XBT)

the bottle to turn upside down. As the bottle is upended, the valves at each end close and a water sample is trapped. This water sample is brought to the surface and analyzed.

The basic temperature measuring device is the reversing thermometer. Usually a Nansen bottle has at least two reversing thermometers attached to it. One of these is a protected thermometer and the other is unprotected, but they both work in essentially the same way. When they are turned upside down, the string of mercury inside each thermometer separates at a constricted region, preventing any subsequent change in temperature reading. In this manner when the thermometers are returned to the surface, their readings report the water temperature at the depth where they were upended. In-place measurements such as these are referred to as *in situ* measurements. The protected thermometer bulb is surrounded by an additional glass chamber so that ambient pressure will not cause an erroneous reading. This chamber is partially filled with mercury to provide adequate heat transfer to the thermometer bulb. The unprotected thermometer does not have this chamber, so that water pressure will cause the thermometer to read somewhat higher than the true value. The difference between the protected and unprotected readings is related to the depth, allowing determination, not only of the water temperature at the sample depth, but also the depth at which the temperature and sample were taken.

In addition to reversing thermometers, another instrument which has come into use in recent years is the thermistor chain. This is a long

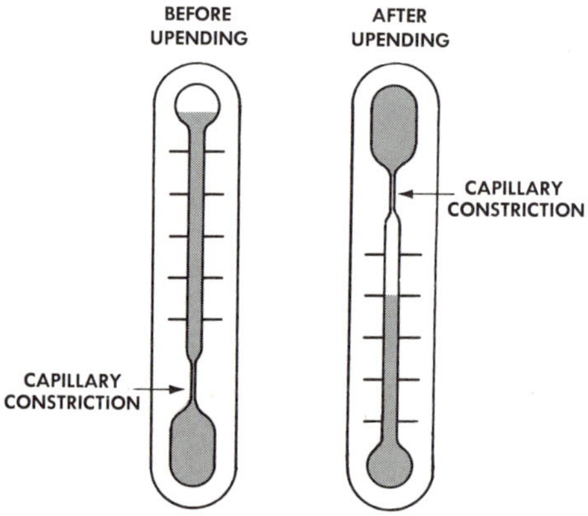

Figure 6-10 The Principle of Operation of the Reversing Thermometer

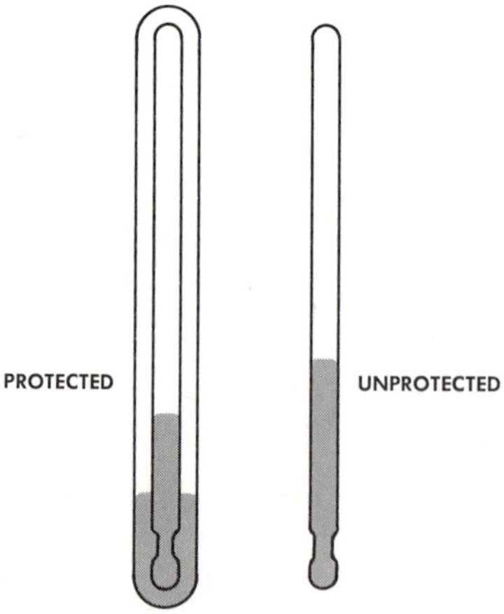

Figure 6-11 Protected and Unprotected Thermometers

Figure 6-12 Nansen Bottle with Reversing Thermometer

cable containing a great number of temperature-sensitive devices called thermistors. A determination of the temperature at perhaps one or two hundred different depths while the chain is being towed behind a vessel allows continuous three-dimensional temperature structure sampling of a relatively large area of the ocean.

605 Temperature Prediction. The temperature structure of the ocean is of more than passing interest to the operational Navy and to commercial fishing fleets. Since the distribution of various fishes is regulated by temperature, and underwater sound is a major tool for detecting both submarines and fish, it is imperative that the effect of the environment on sound transmission be known. A major problem in the environment is the variation of temperature with depth, because temperature causes a great deal of sound-ray bending and affects sound speed more markedly than does any other single factor. It therefore is desirable to predict the depth of the diurnal and seasonal thermoclines.

Figure 6-13 Thermistor Chain Being Used Aboard USS *Marysville* (EPCER-857)

In order to predict environmental factors such as thermocline depths and specific temperature locations, a number of different systems have been developed. The large data input that is required, especially if predictions are desired on a daily basis, is presently being supplied by naval ships, merchant vessels, fishing vessels, satellites, and a few buoys. However, a large buoy system, which will supply data from areas where ships do not normally travel, is being implemented.

As forecasting techniques improve, not only will naval commanders be better able to evaluate the performance of sonar gear, but fishermen, both sport and commercial, will be able significantly to increase their catches. In addition, because the procession of weather systems that crosses our continents derives so much of its energy from the latent heat of vaporization resulting from oceanic evaporation, accurate predictions of surface-water temperature should aid immeasurably in weather forecasting. Thus, a whole spectrum of activities is influenced by knowledge of oceanic temperature structure and its variation with time.

Additional Reading

Defant, A., *Physical Oceanography*, Vol. I, Pergamon Press, 1961.

Dietrich, G., *General Oceanography*, John Wiley & Sons, 1963.

Duxbury, A. C., *The Earth and Its Ocean*, Addison-Wesley Publishing Company, 1971.

Sverdrup, H. V., Johnson, M. W., and Fleming, R. H., *The Oceans*, Prentice-Hall, Inc., 1942.

Weyl, P. K., *Oceanography*, John Wiley & Sons, 1970.

Williams, J., *Oceanography*, Little, Brown & Co., 1962.

CHAPTER SEVEN

The Fluid Behavior

701 Adiabatic Processes. For many purposes the air in the atmosphere and the water in the oceans may be treated similarly because they are both fluids. They both flow and although the atmosphere is much more compressible than the hydrosphere, both will indeed change volume upon the application of a force. This property of compressibility accounts for many of the similarities and, conversely, dissimilarities of the two fluids. Some of the more interesting pressure effects can be observed by first looking at adiabatic processes.

Consider a long thin box having a cross-sectional area of one square centimeter, a length of ten thousand meters, and negligible weight. For demonstration purposes, the box is filled with a compressible gas (air) and placed on its side. Upon examination, the air's temperature is found to be essentially constant throughout the box (Figure 7-1).

Now suppose that the box is stood up on its end. If the temperature is measured within the box, the gas at the bottom is found to have in-

Courtesy: J. A. Perrenoud, Chesapeake, Virginia

creased its temperature while the upper temperature has decreased. At the bottom the density and pressure have increased due to the weight of the fluid column above causing a temperature increase, while the temperature decrease at the top of the box is associated with a decrease in both density and pressure in that area. In general, the average temperature of the box is now greater than it was before.

What is the source of the heat energy increase? When the box is erect, part of the energy utilized in erecting it is now in the form of potential energy or energy of position; however, some has been converted into heat energy, showing up as a change in temperature of the fluid. Note that this change in temperature is associated with the compressibility of the gas since, if the fluid were not compressible, the center of gravity of the box would coincide with the box midpoint in the upright position and the temperatures would not have changed.

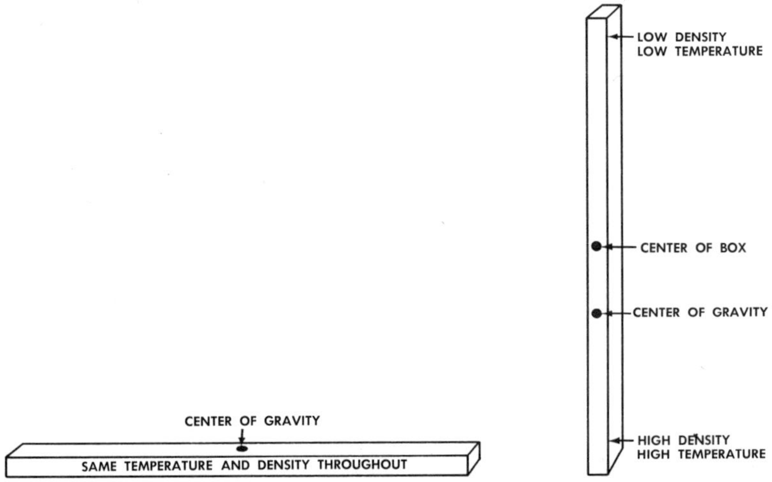

Figure 7-1 The Adiabatic Effect Produced by Placing a Long Box on its End

If mechanical work is done on a fluid without either adding or subtracting heat, the process involved is called an *adiabatic* (Greek, not transferred) *process*. The fluid in the box had work done on it, but no heat was exchanged. The temperature change resulted from mechanical work, not the addition of heat. This is an example of an adiabatic process. In nature most of the processes by which air and water suffer pressure changes by means of vertical movement may be assumed with very little error to be adiabatic.

As a result of the process just described, water that is compressed by the weight above it will be warmed at the ocean bottom. Also air at the earth's surface will warm for the same reason. In the atmosphere warmer air is normally found at the bottom of the atmospheric column; whereas in the hydrosphere the natural temperature of bottom water is usually cold enough to more than offset this adiabatic elevation.

702 Adiabatic Lapse Rate and Temperature Gradient. As would be expected, the amount of temperature increase due to adiabatic heating is directly related to the fluid compressibility. Since air is much more compressible than water, the adiabatic heating in an air column is much more pronounced than that in a water column. If the box were filled first with dry air and then with water, the variation of temperature with distance from the top when the box was raised would be given by the following table.

Table 7-1 Adiabatic Lapse Rate and Temperature Gradient

DEPTH (IN METERS)	DRY AIR TEMPERATURE	WATER TEMPERATURE
Top of box	0°C	0°C
1,000	10°C	0.05°C
4,000	40°C	0.31°C
8,000	80°C	0.91°C
10,000 (Bottom of box)	100°C	1.20°C

Not only is the atmospheric change much greater in magnitude than that in the ocean, but the adiabatic rate of change in temperature is a constant (1°C per 100 meters) in the atmosphere. In the hydrosphere, it varies markedly with pressure (Figure 7-2).

A change in temperature with vertical distance is called a *lapse rate* in the atmosphere (as discussed in Chapter 4), and a *temperature gradient* in the ocean. The adiabatic lapse rate for dry air in the atmosphere is very close to 1°C per 100 meters of altitude. In the ocean it varies from

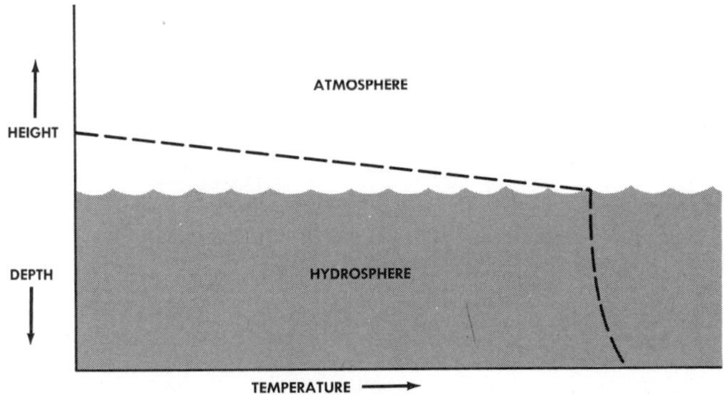

Figure 7-2 Adiabatic Temperature Change in the Atmosphere and the Hydrosphere

about 0.0035°C per 100 meters near the surface to as much as 0.17°C per 100 meters at a depth of 10,000 meters.

There is one further complication in the atmosphere produced by the fact that one of the constituents of air is a varying amount of water vapor. When air is lifted it will expand because of the lessened pressure at altitude resulting in a continual temperature decrease. This mechanical work done by air in expansion is the cause of the adiabatic lapse rate. However, as soon as the dew point is reached (relative humidity = 100 percent) any further work done by the air in expansion is partially the result of the release of the latent heat of vaporization from water vapor as it condenses. In other words, when condensation is taking place the rate at which temperature is decreasing with altitude is not as great because the released latent heat tends to warm the air.

This may be illustrated by assuming a parcel of air containing a fixed amount of heat and a relative humidity well below saturation. If this parcel be lifted, its volume will increase. Since this is an adiabatic process, only the original amount of energy is available to be distributed over a larger volume causing the temperature to decrease. If enough water vapor is added to this same parcel to reach saturation, a slightly different result is obtained when the parcel is lifted. As before, the volume increases and the temperature decreases as a direct result of this expansion. Since saturation was present before lifting took place, any decrease in temperature will produce progressive condensation of the water vapor with the associated release of latent heat. This released latent heat

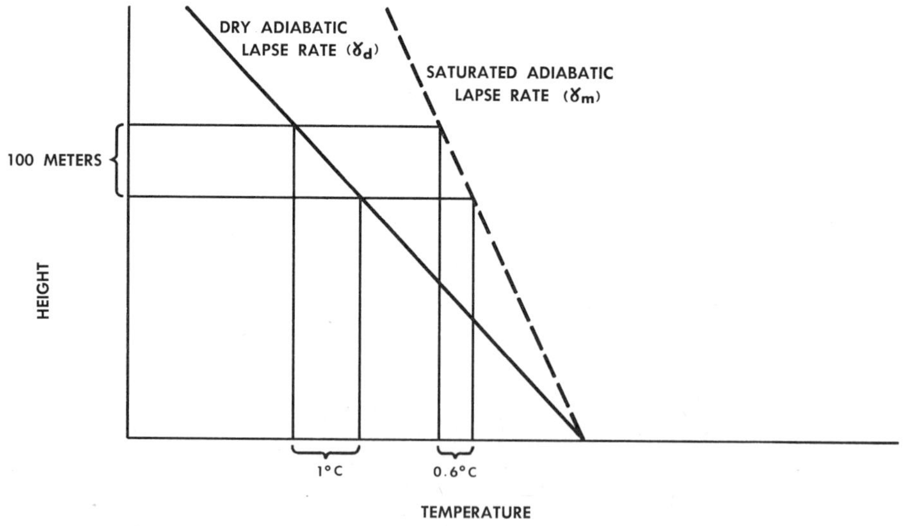

Figure 7-3 Dry and Saturated Adiabatic Lapse Rates

is transferred to the air in the parcel preventing the temperature from decreasing as much as it would if saturation were not present.

To account for the latent heat of vaporization it is necessary to introduce a new lapse rate called the *saturated adiabatic lapse rate.* This saturated (sometimes called moist or wet) adiabatic lapse rate is at all times less than the dry adiabatic lapse rate and varies with the temperature, but on the average, it is about 0.6°C per 100 meters. Meteorologists usually use the symbol γ_d for the dry adiabatic lapse rate and γ_m for the moist adiabatic lapse rate (Figure 7-3).

703 Density. One may very well inquire as to how this variation in temperature produced by the fluid compressibility affects the fluid density. In the ocean there are three parameters, temperature, salinity, and pressure, which affect density. Since density is often determined to five decimal places, in order to make the presentation of these data more clear and to allow for easier computation, the oceanographer introduces a new parameter which he calls sigma (σ). In general, sigma is given by the following:

$$\sigma_{S,T,p} = (\rho_{S,T,p} - 1)1{,}000 \qquad (1)$$

In essence the formula states: to convert density at a particular salinity (S), temperature (T), and pressure (p) to sigma at the same salinity, temperature and pressure, one needs simply drop the 1 and move the decimal point over 3 places to the right. In other words, a water sample having a density of 1.02367 has a sigma of 23.67.

When comparing waters which are found at depths not differing by more than a few hundred meters, it may be assumed that the pressure effect on density is negligible to a first approximation. This allows the introduction of sigma t (σ_t), the sigma associated with the variation of temperature and salinity only.

$$\sigma_t = \sigma_{S,T,o} = (\rho_{S,T,o} - 1)1{,}000 \qquad (2)$$

At the surface, of course, all sigma values are sigma t values. A sample of water taken from a given depth and brought to the surface will have a lower density value at the surface ($p = o$) than at depth. This new density will be associated with sigma t.

704 T-S Diagram. A very convenient method for examining the variation of sigma t with temperature and salinity is the *T-S* curve. Since sigma t is related only to the temperature and salinity, for every temperature-salinity combination there will be associated a unique value of sigma t. A grid of temperature vs salinity may then be constructed and sigma t curves plotted on it. Such a plot is shown in Figure

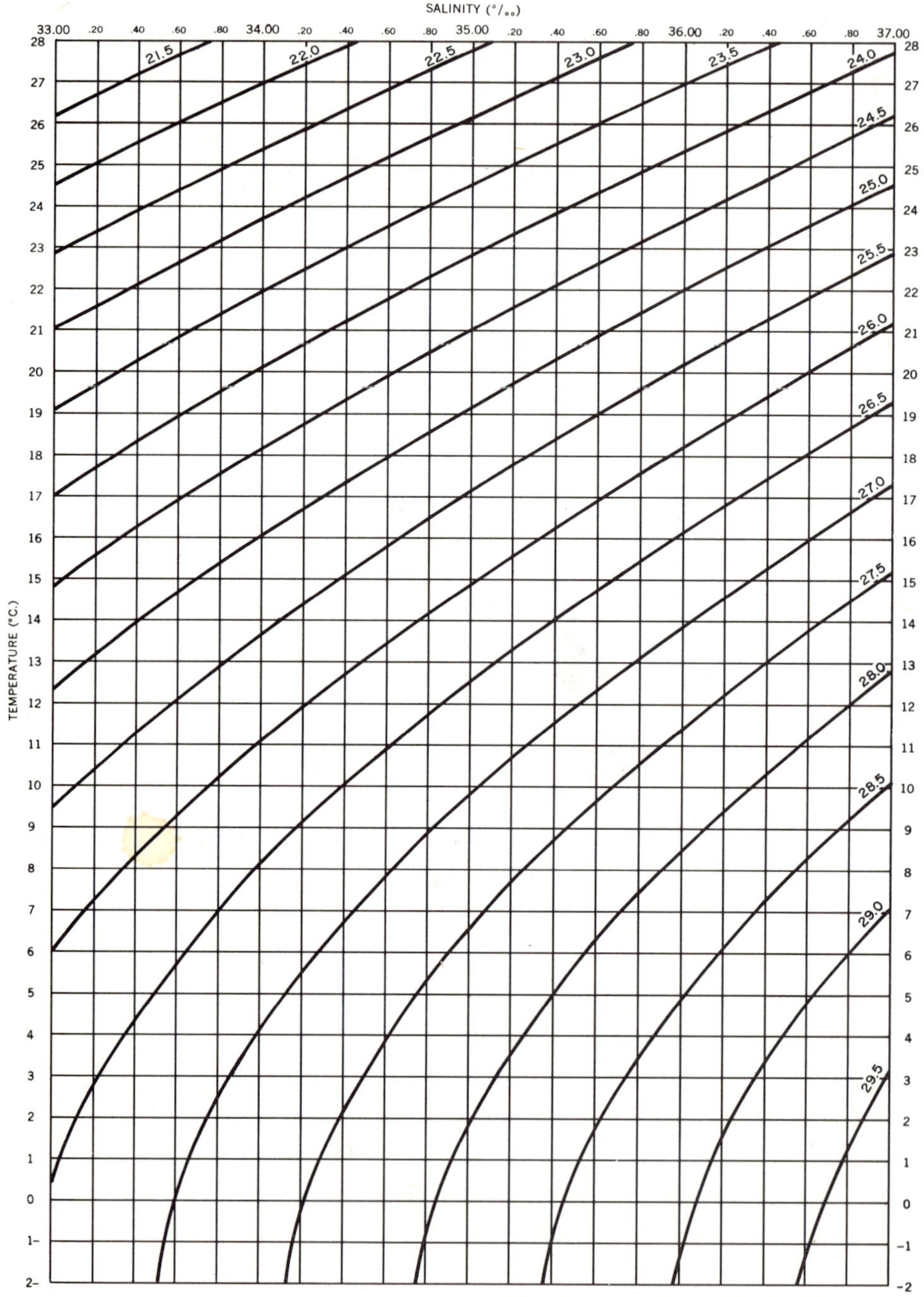

Figure 7-4 The T-S Grid with Sigma *t* Curves

7-4. Note that the same value of sigma *t* may be associated with a high temperature-high salinity combination and a low temperature-low salinity combination, since these two parameters affect density in opposite manners.

705 Salinity Variation. About 99.5% of the water in the oceans has a salinity between 33‰ and 37‰. Actually there are not many processes that produce changes in salinity under natural conditions. The salinity of surface water may be increased by: 1) an excess of evaporation over precipitation or runoff; 2) the formation of ice; or 3) some mixing process that brings higher-salinity waters from below to mix with those at the surface.

Conversely, a decrease in salinity may be produced by: 1) an excess of precipitation or runoff over evaporation; 2) melting ice; or 3) mixing of lower-salinity waters from beneath.

Subsurface waters can change in salinity only by mixing with other water masses.

706 Temperature Variation. Density may also be affected by temperature and, in the upper layers, the major physical processes tending to affect the water temperature are solar radiation, back radiation, evaporation, and transfer of heat to the atmosphere. Once the water sinks, the only process available for changing its temperature is mixing, just as in the case of salinity.

It appears that most water masses acquire their salinity-temperature characteristics at the surface. These are only changed by mixing with other water masses once the water leaves the surface.

707 Temperature and Salinity Effects on Density. In the atmosphere temperature is really the only parameter which predominantly affects the density of the air. Consequently, meteorologists find that density variations in the atmosphere are completely described by discussion of the temperature variation with pressure. It is necessary only to give the actual lapse rate of the atmosphere to describe completely

Table 7-2 Density-Height Variation for Ocean and Atmosphere

SEAWATER				ATMOSPHERE		
HEIGHT/DEPTH Meters	S ‰	T (°C)	DENSITY (gm/cm^3)	PRESSURE (mb)	T (°C)	DENSITY (gm/cm^3)
0	36.40	22.11	1.02527	1,013	15	12.3×10^{-4}
1,000	35.02	5.90	1.03217	905	8.5	11.2×10^{-4}
2,000	34.96	3.43	1.03700	810	2.0	10.25×10^{-4}
3,000	34.95	2.66	1.04156	720	−4.5	9.35×10^{-4}
4,000	34.92	2.31	1.04593	640	−11.0	8.51×10^{-4}

the density variation with altitude. In the atmosphere the temperature variation with altitude changes with location and with time. However, typical lapse rates seem to average about .65°C per 100 meters. If an atmosphere with an average lapse rate of .65°C per 100 meters is contrasted with a typical variation of temperature and salinity with depth in the ocean the density-height values are as given in Table 7-2.

Figure 7-5 A Modern Salinity/Temperature/Depth Sensor Being Lowered Over Side of Coast Guard Oceanographic Vessel USCGC *Evergreen* (WAGO-295)

Note that salinity and temperature data as a function of depth for the ocean, and temperature and pressure data as a function of altitude for the atmosphere are sufficient to calculate densities for both media. For this set of data taken in the Atlantic Ocean the density varies from 1.02527 gm/cm^3 at the surface to 1.04593 gm/cm^3 at a depth of 4,000 meters. In the atmosphere at this particular station, the density varies from 0.00123 grams per cubic centimeter at the surface to 0.000851* grams per cubic centimeter at an altitude of 4,000 meters. In addition to the ocean water being about 800 times as dense as the atmosphere, note that the relative change in density over the same vertical distance is very much greater in the atmosphere than it is in the hydrosphere. For this oceanic station the density at a depth of 4,000 meters is 2 percent greater than that at the surface. In the atmosphere, however, at an altitude of 4,000 meters the density is 31 percent less than that at the surface.

Consequently, it may readily be seen that both the temperature and pressure effects on density are very much smaller in the ocean than in the atmosphere. Therefore unless densities representing great depth differences are compared the effect of pressure on density may be neglected in the ocean. In the atmosphere, this is far from being the case since small changes in altitude markedly affect the density of air.

708 Stability. If the variation of density with altitude and depth is such that the heavier fluid is beneath a lighter fluid, the fluid column is said to be *positively stable* (stable). Conversely, if the lighter fluid is on the bottom and the heavier fluid is on top, the column is said to be *negatively stable* (unstable). In the ocean one may examine for stability by plotting the actual temperature and salinity values at different depths on the *T-S* grid discussed previously. The data are plotted and the points are connected *in order of increasing depth.*

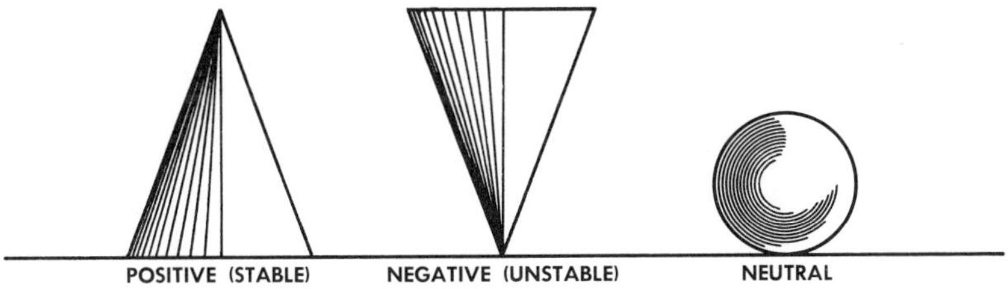

Figure 7-6 Examples of Stability

* The meteorologist has no density parameter comparable to the σ of the sea. It is probably because the parameter of atmospheric density is difficult to measure and is normally considered only in terms of two atmospheric density determinants, pressure and temperature.

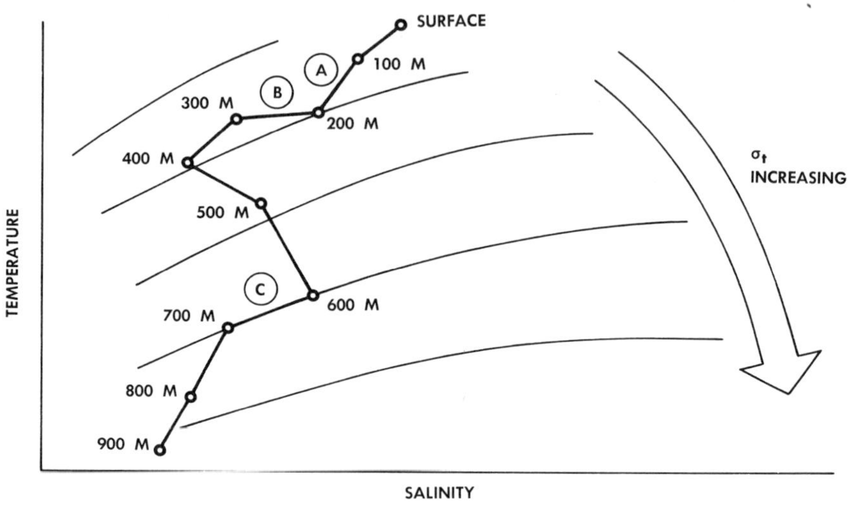

Figure 7-7 Use of the T-S Diagram to Determine Stability

Figure 7-7 shows plots of the variation of salinity and temperature with depth for a station in the North Pacific Ocean. From these data the temperature and salinity values at each depth of interest are obtained and plotted on the *T-S* grid, so that each point on the *T-S* curve represents the temperature and salinity conditions at a single depth. Since σ_t curves form a permanent part of the background grid, the density-depth variation and, thus, the stability condition of various portions of the water column may easily be determined.

If points representing increasing depth are associated with increasing values of σ_t, a stable condition is present, as is shown by portion *A*. However, if a portion of the water column has a density distribution such that the water beneath is less dense than that above, such as in region *B* in the figure, negative stability is said to be present. A third stability condition resides between the stable and the unstable condition such that the density does not change with depth. This is called *neutral stability*. For this particular station, between the depths of 600 meters and 700 meters, the *T-S* curve is parallel to the sigma *t* curves, so that region *C* of Figure 7-7 represents neutral stability.

In the atmosphere the determination of stability is somewhat more complicated than in the hydrosphere. However, the basic principle is the same: if the fluid below is more dense (colder) than the fluid above, the condition is a stable one, and if the fluid below is less dense (warmer) than that above, the condition is an unstable one. Since a small change in pressure produces such a large change in density in the atmosphere, the effect of pressure cannot be neglected. When comparing the density of one air parcel to that of another *the comparison must be made at the same pressure level.* To make a comparison at the same pressure level, one must assume hypothetical vertical motions wherein adiabatic processes are observed. If the air is not saturated, any increase in altitude must be done at a dry adiabatic rate such that the temperature decreases one degree Celsius per 100 meters of altitude. If the air is saturated, the temperature must decrease at a saturated adiabatic rate.

For example, if it is desired to determine the stability of an existing air column, the air at the base of the column is hypothetically raised to the top. When the air reaches the top of the column, having experienced an adiabatic transformation as indicated, and if its temperature is less than that which is already there, the column is stable. This is true since a lower temperature is associated with a higher density. However, if its temperature were greater than that of the air at the top of the column, it would be less dense and thus unstable.

Actually all one is doing is comparing the actual vertical variation of temperature with the adiabatic lapse rate. This is shown in Figure 7-8. The dry adiabatic lapse rate, characterized by a decrease in temperature of 1°C for every 100 meters increase in height is represented by the solid line, while the dotted line indicates the actual variation of temperature with altitude. Case 1 represents a lapse rate which is less than that of the adiabatic lapse rate (the temperature changes less than 1°C in 100 meters of altitude). It may be seen that this air column is stable, because any parcel of air lifted adiabatically from the lower portions of the column will be lifted in such a manner that its temperature will always be less than that of the actual air represented by the dotted line. At any altitude the result will be that the lifted air is

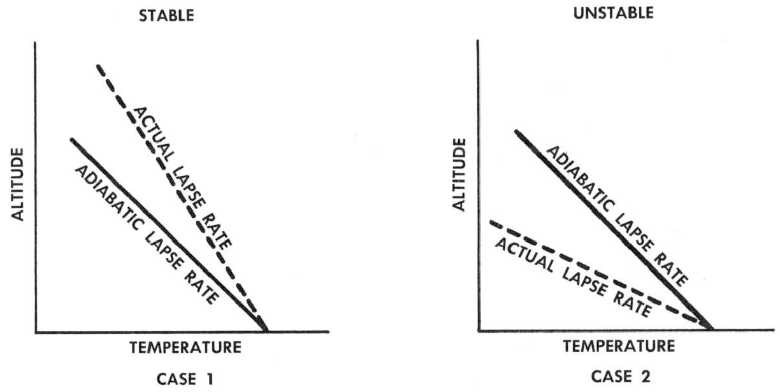

Figure 7-8 Stability in the Atmosphere as Indicated by Comparison of Actual and Adiabatic Lapse Rates

more dense and falls back to its original position. *A column of air having a lapse rate which is less than the adiabatic lapse rate is said to be stable.*

On the other hand, *if the actual lapse rate of a column of air* (indicated in Case 2) *is greater than that of the adiabatic rate* (the temperature changes more than 1°C in 100 meters of height) *an unstable condition results*. When a parcel of air is lifted along the solid line (the dry adiabatic lapse rate), at any altitude it will have a temperature greater than that actually present, and therefore will be less dense tending to rise rather than fall back to its initial position.

If the actual lapse rate is identical with the adiabatic rate, the resulting state is called neutral stability. No matter where the parcel of air is placed in the atmospheric column it will have the same density as the surroundings and will have a tendency to neither rise nor fall. As a matter of fact, this condition is very rarely seen in the atmosphere, but is mentioned for the sake of completeness.

These three conditions of stability have been discussed in terms of comparison with the dry adiabatic lapse rate. However, if the air is saturated, the comparison for determining stability must be made with the saturated adiabatic lapse rate and the same conditions hold.

In Figure 7-9 both the dry adiabatic lapse rate (γ_d) and the saturated adiabatic lapse rate (γ_m) are plotted. Note that the saturated adiabatic lapse rate is somewhat less than the dry, since it represents a smaller

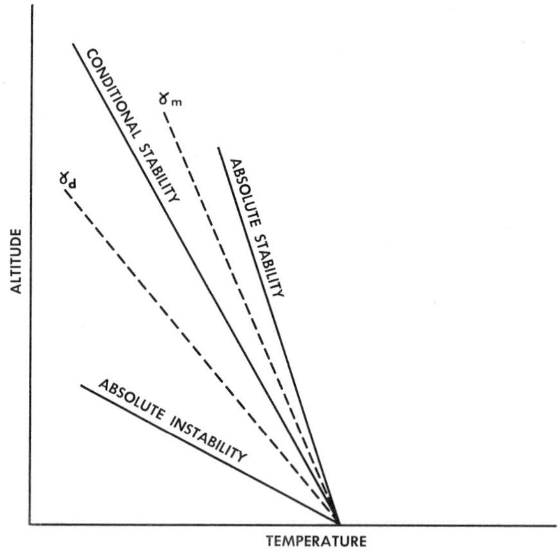

Figure 7-9 Various Types of Atmospheric Stability (The Solid Lines Represent Typical Radiosonde Traces.)

change in temperature for a given change in altitude. If the existing lapse rate falls somewhere between the dry and moist adiabatic lapse rates, this condition is called *conditional stability*. Stability is conditioned upon the saturated or unsaturated state of the air. For example, a parcel of air having a relative humidity of less than 100 percent and a lapse rate in between the moist adiabatic lapse rate and the dry, is caused to be lifted. This air mass cools as it is lifted and it is stable. However, as soon as the dew point is reached and water vapor starts to condense, any further lifting will necessitate a comparison, for stability-determining purposes, to be made with the wet adiabatic lapse rate. It may be seen that the air mass is now unstable so that a conditionally unstable air mass is stable until condensation is reached and unstable thereafter.

If, on the other hand, the actual lapse rate is less than both the saturated lapse rate and the dry lapse rate, this air will always be stable. This is called *absolute stability*.

If the lapse rate is greater than both the dry and the saturated adiabatic lapse rate, the air will always be unstable. This condition is called *absolute instability*.

709 Atmospheric Stability Summary. In summary the atmospheric stability conditions are as follows:

1. *Stable air* results when the lapse rate is less than the dry adiabatic lapse rate if the relative humidity is less than 100 percent, or less than the saturated adiabatic lapse rate if the relative humidity is 100 percent.
2. *Unstable air* results when the lapse rate is greater than the dry adiabatic lapse rate if the relative humidity is less than 100 percent, or greater than the moist adiabatic lapse rate if the relative humidity is 100 percent.
3. *Neutral stability* results when the lapse rate is equal to the dry adiabatic lapse rate for dry air or the saturated adiabatic lapse rate for saturated air.
4. *Conditional stability* results when the actual lapse rate is in between the wet and the dry adiabatic lapse rates.
5. *Absolute stability* results when the actual lapse rate is less than the saturated adiabatic lapse rate.
6. *Absolute instability* results when the actual lapse rate is greater than the dry adiabatic lapse rate.

710 Convergence and Divergence. One of the major results of instability is vertical motion, and whenever vertical motion is present

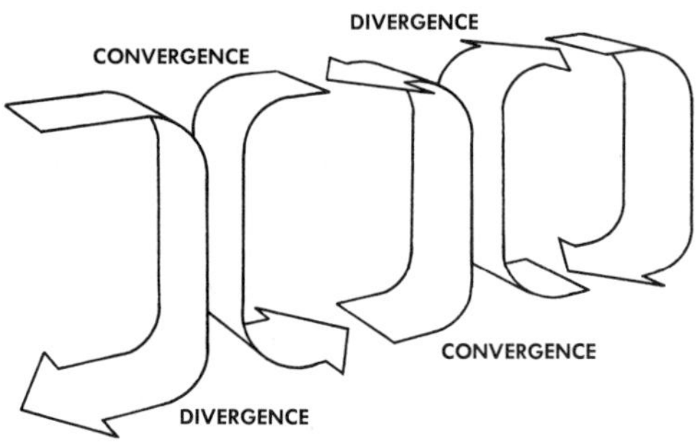

Figure 7-10 Convergence and Divergence Produced by Vertical Motion

there must be either convergence or divergence in a horizontal plane. In the case of convergence the fluid must go somewhere. In divergence the fluid must be supplied from somewhere. Figure 7-10 indicates both a convergent region in the atmosphere at the surface producing vertical motion upwards, and at some higher elevation, a divergent region associated with this vertical motion. Similarly, on the other side of the convection cell there is a convergent region at an altitude associated with subsiding air and at the ground there is a region of divergence. In general, *any* fluid vertical motion is associated with either convergence or divergence since the atmosphere and the hydrosphere are both three-dimensional in nature.

If two fluid masses tend to move in a direction so that they will come together, the region of confluence is called a convergent region, and this process is called *convergence*. This may result from motions which are oppositely directed, or it may result from motions in the same direction at different speeds.

Figure 7-11 Convergence Due to Direction Difference

In Figure 7-11 one can see that convergence results from the two motions indicated by the arrows coming together. Figure 7-12 also shows a region of convergence, however, the motion on the right is less rapid than that on the left. Now fluid behind will catch up to the fluid ahead resulting in convergence.

Figure 7-12 Convergence Due to Speed Difference

In a similar manner, one may discuss the phenomenon of divergence. *Divergence* is the converse of convergence; that is, when fluids tend to flow apart there is divergence. Again this may result from two fluid masses flowing in opposite directions as indicated in Figure 7-13 or

Figure 7-13 Divergence Due to Direction Difference

flowing in the same directions, but at different speeds as indicated in Figure 7-14.

Figure 7-14 Divergence Due to Speed Difference

Divergence and convergence are commonly occurring phenomena and shall be met again, since they form an important part of both the atmospheric and hydrospheric circulation patterns.

Additional Reading

Barry, R. G., and Chorley, R. J., *Atmosphere, Weather, and Climate*, Holt, Rinehart, and Winston, 1970.

Blair, T. A. and Fite, R. C., *Weather Elements*, 5th ed., Prentice-Hall, Inc., 1965.

Byers, H. R., *General Meteorology*, 3rd ed., McGraw-Hill, Inc., 1959.

Dietrich, G., *General Oceanography*, John Wiley & Sons, 1963.

Donn, W. L., *Meteorology*, 3rd ed., McGraw-Hill, Inc., 1965.

Duxbury, A. C. *The Earth and Its Ocean*, Addison-Wesley Publishing Co., Inc., 1971.

Neumann, G. and Pierson, W. J., *Principles of Physical Oceanography*, Prentice-Hall, Inc., 1966.

Petterson, S., *Introduction to Meteorology*, McGraw-Hill, Inc., 1958.

Riehl, H., *Introduction to the Atmosphere*, McGraw-Hill, Inc., 1965.

Sverdrup, H. V., Johnson, M. W., and Fleming, R. H., *The Oceans*, Prentice-Hall, Inc., 1942.

Von Arx, W., *Introduction to Physical Oceanography*, Addison-Wesley Publishing Co., Inc., 1962.

Williams, J., *Oceanography*, Little, Brown, & Co., 1962.

XII
XI
X
IX
26
25
24
22
21
20
19

CHAPTER EIGHT

Light and Sound Energy Transmission

801 Waves. This chapter examines the effect of the atmosphere and the hydrosphere on the passage of radio waves, light waves, and sound waves. It is only logical to lump these three forms of energy together since they have many characteristics in common.

All three involve an oscillatory motion. In the case of sound waves, the medium itself, whether it be air or water, is moving back and forth as the wave goes by, whereas in the case of light and radio waves, the associated magnetic and electric fields both vary with time.

In addition to the similarity involving oscillatory motion, all wave energy is subject to the effects of absorption, scattering, reflection, refraction, and diffraction.

There will be two major areas of concern. The energy losses within the medium will be examined, and then the variation in the speed of propagation due to the effects of a changing medium will be explored.

Rubber sonar dome of frigate USS Willis A. Lee

802 Basic Definitions. Every wave may be specified by three parameters. These are the *wavelength* (L),* the *period* (T), and the *wave speed* (c). On a plot of electric-field strength (electromagnetic wave) or of pressure (sound wave) as a function of time, the *wavelength* is the distance between two points on the curve having the same phase. This is illustrated in Figure 8-1.

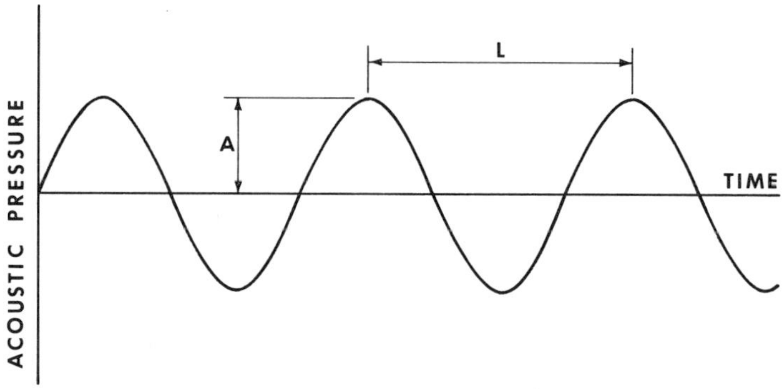

Figure 8-1 Plot of a Sound Wave

The time required for one wavelength to pass a given point in space is called the *period.* The reciprocal of the period is the *frequency* (f), usually expressed as a number of cycles per unit time.

If the two parameters of wavelength and period are combined, the *wave,* or *phase, speed* (c) is given by the ratio of the wavelength (L) to the period (T), that is:

$$c = \frac{L}{T}. \tag{1}$$

Frequency will be specified in terms of *hertz* (Hz), wherein one hertz is equal to one cycle per second.

Table 8-1 Multiple Prefixes for Wavelengths

MULTIPLE	PREFIX	SYMBOL
10^6	mega	M
10^3	kilo	k
10^{-1}	deci	d
10^{-2}	centi	c
10^{-3}	milli	m
10^{-6}	micro	μ

* Note that this is the same parameter denoted by λ in Chapter 5.

It will also be convenient to discuss the *amplitude* (A) of the wave, shown as the maximum displacement of the curve in Figure 8-1. The energy carried by a wave is proportional to the square of the amplitude, and a term called *intensity* (I) is a function of the wave energy. By intensity is meant the energy passing through a perpendicular unit area per unit time. Thus the units of intensity are energy per unit area per unit time. In the measurement of wave length the metric system will be used with the associated prefixes.

803 Light Losses. Generally speaking, light is defined as that portion of the electromagnetic spectrum to which the human eye is sensitive. An *average observer* can see light having wavelengths somewhere between 0.35 micrometers and 0.75 micrometers which covers the entire spectrum of colors from deep violet to deep red, with the blues, greens, and yellows being interspersed.

As light travels through any medium it is subject to two basic losses. These are absorption and scattering. *Absorption* is conversion of light energy into heat, while *scattering* is essentially a change in direction of light energy produced by particulate matter in suspension. This particulate matter may be anything from air molecules and rain drops in the atmosphere to suspended dirt particles in the hydrosphere. If there are no particles in suspension, and if the scattering effects of the individual molecules of the medium are neglected, absorption may be assumed as the only loss present.

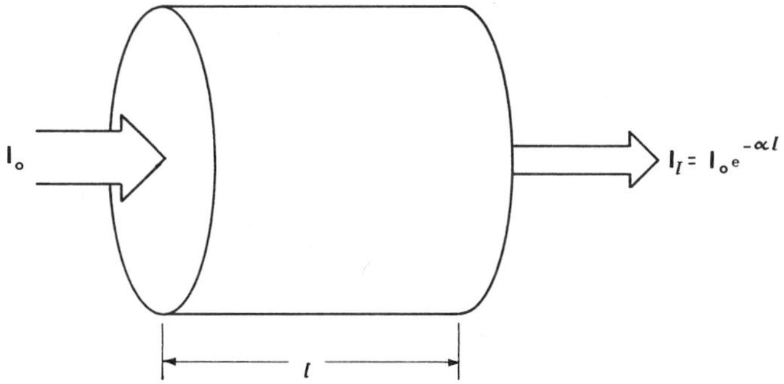

Figure 8-2 Lambert's Law

Suppose a section of medium having a length l is illuminated with light of intensity I_0. If the medium is optically homogeneous the fraction of light energy lost per unit length will be a constant. If I_l is the emergent light, $(I_0 - I_l)/I_0$ is the fraction of light lost in a section of medium having a length l, and $\Delta I/I$ will be the fraction of light lost in the incremental distance Δl. So that with an optically homogeneous medium

$$\frac{\frac{\Delta I}{I}}{\Delta l} = -\alpha$$

where α is the constant previously referred to and the minus sign results from the fact that this process is an energy loss. In differential form this becomes

$$\frac{dI}{I} = -\alpha dl$$

which when integrated between the limits of $l - 0$ and $l = l$ results in Lambert's Law:

$$\ln I_l - \ln I_0 = -\alpha l \qquad (2)$$

or, in exponential form:

$$\frac{I_l}{I_0} = e^{-\alpha l} \qquad (3)$$

where: e is the natural logarithm base (2.718 . . .)
α is the absorption coefficient
l is the length of the medium
I_0 is the incident intensity, and
I_l is the intensity after traversing a path equal to l.

804 The Characteristic Absorption Length. The absorption coefficient (α) has the units of reciprocal length; in oceanography this is usually per meter, and in meteorology, per kilometer. Alpha varies in value from 0 for a perfectly transparent medium, to infinity for a perfectly opaque one. Since this parameter is somewhat difficult to grasp intuitively, it is often desirable to introduce a new parameter called the *characteristic absorption length* (£).*

The *characteristic absorption length* is defined as the distance light must travel to be reduced in magnitude to $1/e$ of its initial value

$$\left(I_{£} = \frac{I_0}{e}\right).$$

For this to be true,

$$\frac{I_{£}}{I_0} = \frac{1}{e} = e^{-1} = e^{-\alpha £}$$

or $-1 = -\alpha£$ and

$$£ = 1/\alpha \qquad (4)$$

* This symbol is commonly referred to as Libra.

Since $1/e$ is about 0.368, the intensity of the light will be reduced to 36.8 percent of its initial value when the light has passed through a thickness of £ in length.

In the atmosphere, absorption is usually a minor loss for light energy. However, in the hydrosphere, absorption may be a major loss. Absorption in the ocean is not only large, but it is related to the wavelength of the light involved. Figure 8-3 shows the variation of the characteristic absorption length with wavelength. Note that water is relatively opaque to both violet and red light, especially red, while it is relatively transparent in the blue-green region. This has the effect of changing the character of sunlight as it passes through water. Photographs taken underwater using natural light are almost monochromatic in the blue-green due to the fact that sunlight loses all of its red and violet components after a relatively short penetration. For blue-green light in absolutely clear water, £ is about 100 meters.

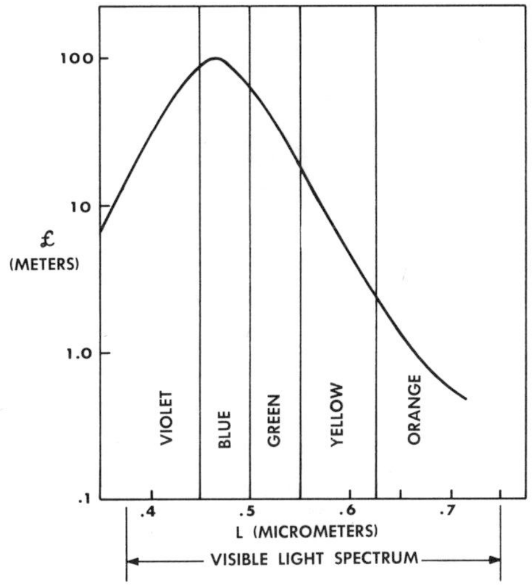

Figure 8-3 Characteristic Absorption Length for Water

Even though the attenuation of light in water is quite high, perceptible amounts of sunlight may be found at great depths. This is due in no small part to the extreme sensitivity of the human eye. Most people can read a newspaper under conditions of bright moonlight which is about 10^{-6} the illumination of a clear summer day. Conditions equivalent to those of moonlight may be experienced at depths greater than 200 meters, while plants will not normally have enough light to grow at depths greater than 80 meters in the same water.

It should be noted that the light-absorptive properties of water are not affected by the dissolved salts. Distilled water and filtered sea water both have the same light-absorption properties.

805 Light Scattering. Normally, in the natural environment, absorption is small compared to the losses incurred by scattering. Both the atmosphere and the hydrosphere usually contain large amounts of particulate matter in suspension, producing large amounts of scattering. In the atmosphere scattering is most pronounced in the blue region of the spectrum since most of the particles are quite small. This is an example of *Rayleigh scattering* which gives rise to the blue color of the sky. The major portion of the blue light from the sun is scattered out of the main beam and reaches the surface of the earth as blue skylight.

Lambert's law, given previously for the process of absorption, may also be utilized for describing scattering losses. The only change necessary is to substitute a scattering coefficient (σ) for the absorption coefficient (α). The scattering-loss law is then:

$$\left(\frac{I_l}{I_0}\right)_{\text{scattering}} = e^{-\sigma l} \qquad (5)$$

Consequently the concept of the characteristic length may be used again, so that the characteristic scattering length ($\pounds_s$) is given by:

$$\pounds_s = 1/\sigma \qquad (6)$$

If both absorption and scattering are present, the total loss is called attenuation, and the characteristic attenuation length ($\pounds_a$) is:

$$\pounds_a = \frac{1}{\sigma + \alpha}, \quad \text{since} \qquad (7)$$

$$\left(\frac{I_l}{I_0}\right)_{\text{attenuation}} = e^{-(\sigma + \alpha)l} \qquad (8)$$

Table 8-2 gives some idea as to the characteristic scattering lengths of pure air. Note how much smaller these numbers are for blue and violet light than for red light.

Table 8-2 Characteristic Scattering Lengths for Dry Air

	ULTRAVIOLET	BLUE	GREEN	RED		INFRARED
L (micrometers)	0.3	0.4	0.5	0.6	0.7	0.8
$\pounds_s$ (kilometers)	6.77	22.9	57.1	120	214	385

It would appear that any long-distance use of light for communication through the atmosphere would best use red or infrared wavelengths. Since most present-day lasers operate best in this range of wavelengths, much greater utilization can be expected in the very near future.

The major effect of this high amount of scattering of blue and violet light is to cause a change in the character of sunlight by the time it reaches sea level. This change in character has the effect of diminishing the peak of solar radiation, which is at about 0.487 micrometers, and making the amount of solar radiation available at sea level just about constant between 0.5 and 0.7 micrometers.

The same effects are present in the hydrosphere; one of the reasons for oceans being blue in color is the scattering of blue light by water particles. In general, the scattering increases as the number of particles in suspension increases. However, since water is more dense than air, it is possible for particles somewhat larger in size to remain in suspension than in the atmosphere. For this reason scattering may take on a somewhat different character, with scattering of red light often being as great, if not greater than, that of blue light.

806 Visibility. Scattering in both the atmosphere and the hydrosphere is, of course, very much affected by the *turbidity* (amount of foreign material present), and in one form or another is the major deterrent to visibility. In the atmosphere this scattering is produced by haze, fog, or smoke present in the air. Usually when the atmosphere is stable the air next to the ground is apt to retain any pollutant which is introduced. Smoke will not be carried aloft, but may contaminate the lower layers to the extent of making the air relatively opaque. For example, smog occurs when automobile exhaust and other pollutants are concentrated so that not only does visibility become impaired, but also the atmosphere becomes somewhat noxious. In this situation, any instability in the atmosphere would tend to disperse the pollutants through a larger volume and increase the visibility. Consequently, clear air and improved visibility are usually associated with unstable air columns. This is not always true since an unstable atmosphere may be accompanied by high winds tending to keep large amounts of dust or spray in suspension.

In the hydrosphere also, foreign material tends to decrease visibility. In general, there are three causes of changes in turbidity: *erosion*, *presence of plankton*, and *mixing*. Erosion is involved with the transport of terrestrial sediments from the land areas out to sea by means of rivers and currents. Living plankton will also tend to increase the turbidity since these creatures will decrease the visibility in much the same manner as inorganic materials. And thirdly, material may be placed in the water from the bottom itself. If the water is shallow, the bottom may be stirred up by mechanical means, both natural and man-made.

807 Ocean Color. Besides lowering visibility, the particulate matter in ocean water contributes to its color. It was indicated previously that a portion of the blue color of the ocean is due to scattering by water molecules. Some of water's color is undoubtedly due to suspended material in the water. Brown water is usually caused by mud in suspension. Water having a greenish cast probably has a large population of phytoplankton. As a matter of fact, blue has been called the desert color of the sea, the color of lifeless water.

In addition to scattering and the suspended materials, there are probably two other factors tending to produce sea-water color. These two are: the dissolved materials present, usually having a yellowish hue, and a reflection of the sky color.

808 Transparency Measurements. Generally speaking atmospheric visibility ranges are estimated by an observer being able to distinguish the details of a fixed object at a known range. Lacking this, the judgement of a qualified observer is required. In the hydrosphere, on the other hand, there are one or two fairly commonly used devices for measuring transparency. One of these is the *Secchi disc*, a white disc about 30 cms in diameter, which is lowered in the water until it just disappears from view. This depth of disappearance is related to the transparency. Also in fairly common use are photoelectric devices called *beam transmittance meters* or *alpha meters* which utilize a light source and a photocell to measure the amount of light loss within a specified water path length.

809 Refractive Index. The speed with which light travels through the atmosphere is very close to that in free space, i.e. about 3×10^8 meters per second. In the ocean, light travels with a somewhat diminished speed, approximately $\frac{3}{4}$ that in free space. The ratio of the speed of light in vacuum to that with which light travels through a given medium is called the *index of refraction*. In the atmosphere the combined effects of pressure, temperature, and water-vapor content produce a range in refractive index values between 1.000250 to 1.000450; a spread usually considered negligible for light, but not for radio frequencies (see sections 812 and 813).

Within the ocean the variation in index of refraction is also quite small. The index of refraction is a function of pressure, temperature, and salinity, varying in the same manner as density. At a chlorinity of 20‰

and a temperature of 0°C the index of refraction is 1.341, while at a chlorinity of 5‰ and a temperature of 25°C, the index of refraction is 1.334. Certainly this is a much larger range than is found in the atmosphere, but for practical purposes it is of little, if any, consequence.*

However, when an observer looks from the atmosphere into the hydrosphere, the change in refractive index is extremely large and here there are important ramifications. The first of these is a change in the apparent size of objects so that they appear somewhat larger. This change in size is greatest around the edges of one's field of view. In other words, objects appear to be larger, and are somewhat distorted. Since objects on the periphery of the field are larger than those near the center, a fish swimming by a viewing port will appear to speed up as he reaches the edge of the visible field. Attempts have been made to correct this distortion for scuba divers by having lenses built into the viewing mask, and for underwater cameras by the use of special lenses. Correction of a viewing port is somewhat more difficult for a submersible, since it would require the viewer to remain in exactly the same location while looking through the corrected window.

Another aspect of the difference in index of refraction between water and air is the fact that sunlight will be reflected from the air-sea interface. This tends to reduce the depth at which submerged objects may be detected as viewed from aircraft. In some cases, due to the glitter and glare produced by small capillary waves and with the sun at the proper angle, it is impossible to see into the ocean at all from a vantage point above the surface. Polarized viewers aid in the solution of this problem.

810 Radio Wave Losses. Farther out along the electromagnetic spectrum, in the direction of increasing wavelength, is the region of radio waves. The portions of the spectrum commonly utilized for communications (radio) and location and tracking of objects (radar) have wavelengths roughly between 1 centimeter and 20 kilometers. Within the atmosphere, absorption of these wavelengths is usually quite small. Within the hydrosphere, the absorption of radio waves is extremely high. It turns out that the absorption is so high that with one major exception (VLF), radio waves cannot be used below the surface of the sea.

In order to get some numerical indication of how high the absorption is, the characteristic absorption length for typical radio waves may be calculated by utilizing a relationship derived from electrical theory. This reduces to the following for sea water of average salinity and temperature:

* The variation of index of refraction is sometimes used as an indicator for the determination of sea-water salinity.

$$\pounds = \frac{226}{\sqrt{f}} \tag{9}$$

where £ is the characteristic absorption length in meters, and f is the frequency in hertz.

At a frequency of 100 MHz (the middle of the commercial TV band) the characteristic absorption length may be calculated to be 2.26 centimeters. Most radar frequencies are quite a bit higher than this and, consequently, would have even shorter characteristic absorption lengths. A somewhat lower frequency on the other hand, that of 1 MHz (the middle of the AM broadcast band) has a characteristic absorption length of about 23 cms. If the frequency is lowered to the very low frequency (VLF) band, 18 kHz, for example, the characteristic absorption length is calculated to be 1.69 meters. The U.S. Navy utilizes this phenomenon by using high-powered VLF transmitters for communications with submerged submarines. Typically, frequencies around 18 kHz are used, since it is only at these low frequencies that any significant amount of electromagnetic energy penetrates to any reasonable depth. Due to the fact that the hydrosphere is opaque to all except very low radio frequencies, all succeeding remarks about radio waves will be confined to the effects of the atmosphere.

Although there is some absorption of radio waves by all the gases comprising the atmosphere, a major portion of any attenuation which accrues is produced by water in the atmosphere, either in the form of gas, liquid, or solid. Typical characteristic lengths for a marine atmosphere of average relative humidity are somewhere around 300 kilometers for 10,000 MHz radar and 600 kilometers for radar in the 3,000 MHz band. In addition to an increase in absorption produced by an increased amount of water vapor, there is also a decrease in absorption with a decrease in frequency; the higher frequencies are more affected by the water vapor than the lower frequencies. For frequencies normally used in communications, absorption losses are so low that they are usually neglected.

811 Effects of Rain. When the water vapor condenses, the attenuation losses increase due to scattering by the individual droplets of water or ice crystals. In the case of water droplets, measurements have been made which indicate that for rain accumulating at a rate of about 1 centimeter per hour, the characteristic length is reduced to about 14 kilometers for 10,000 MHz radar. As the rainfall rate increases, the characteristic length will become even shorter.

An interesting application of this is the effect of a frontal passage (see Chapter 16 concerning fronts) on the maximum radar ranges to be expected. The major reduction in radar range by the front is caused by the rain associated with the front.

812 The Ionosphere in Communications. The major environmental effect on the transmission of radio wave energy is that of refraction. This refraction takes two forms; one occurring in the ionosphere and the other occurring in the troposphere. Refraction produced by the ionosphere was utilized by Marconi in 1901, when he successfully transmitted a message across the wide span of the Atlantic Ocean.

In Chapter 4 it was seen that above the stratosphere is a region of charged particles called the ionosphere, composed mainly of ions produced by the action of incoming solar radiation. This region of high ion density tends to markedly change the index of refraction of radio waves below a certain frequency. It has such a large effect that radio waves of frequencies below about 30 MHz are bent back down toward the earth if the incident angle within this layer is not too great. In this manner radio waves may be refracted back down toward the earth, and long ranges may be obtained by the deflection of this layer along with multiple reflections from the earth's surface as indicated in Figure 4-2. Because solar radiation disappears during the night, the ionosphere height changes. With this nocturnal increase of height, the effective range of most radio waves is enhanced. This effect is well known to commercial broadcasters who are required, in the main, to diminish their power after sunset. As a matter of fact, many stations are required to go off the air after sunset because of possible interference with other stations at great distances on similar frequencies.

At night, for frequencies less than about three MHz, one can expect about 10 times as great a range as during the day. For frequencies greater than three MHz, a less pronounced range increase is observed, averaging about two or three times greater at night than during the day.

Since the refraction produced in the ionosphere is greater for lower frequencies, and the absorption of radio energy in the ionosphere increases as the frequency increases, above about 30 MHz the ionospheric refractive effect is of negligible importance.

Just as daytime conditions are associated with thicker ionization layers than those at night, a similar increase is produced during the summer months when there is more effective solar radiation than during the winter months.

In addition to these periodic, essentially predictable fluctuations, there are aperiodic fluctuations in the composition of the ionosphere mainly produced by magnetic storms within the ionosphere itself. During these periods of magnetic storms, which may last from a few hours to several days, radio transmission is typically quite poor within the

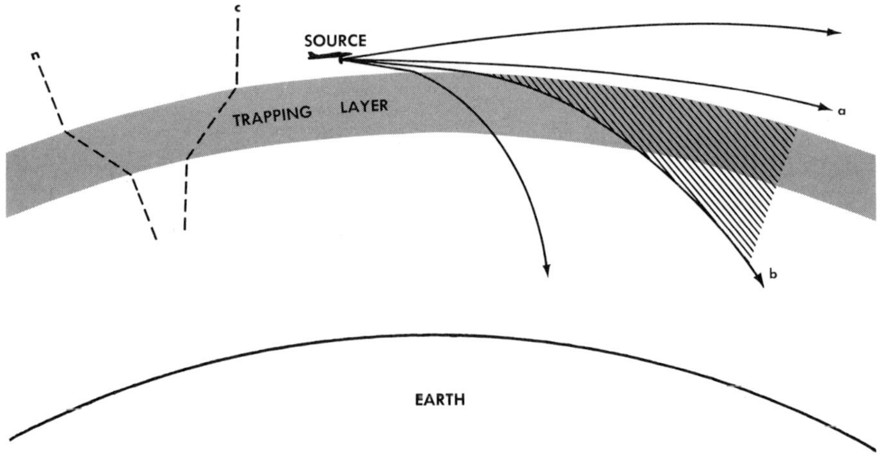

Figure 8-4 The Radar Hole (Hatched Area)

bands normally using the ionosphere. This is due to the rather marked expansion of the ionospheric thickness and the accompanying increased absorption of radio waves.

813 Refraction in the Troposphere. Within the troposphere the index of refraction changes by small amounts related to pressure, temperature, and water-vapor content. An increase in either pressure or water-vapor content tends to increase the index of refraction, whereas an increase in temperature will decrease the index of refraction. Normally the refractive index decreases with altitude since the pressure and water-vapor content are decreasing, and they apparently override the effect of decreasing temperature. However, if for one reason or another there are layers having lower than normal water-vapor content or temperature inversions, the variation of refractive index with height is changed from the common situation. When this occurs, aircraft radar usage becomes somewhat less dependable.

It is possible, for example, for a marked decrease in refractive index with height to produce what is called a *trapping layer*. Associated with these trapping layers are *radar holes* or areas where radar signals will not penetrate. In this situation one portion of the radar beam is caused to bend away from the earth's surface leaving a hole in the area being searched. Figure 8-4 is a schematic representation of the radar hole. The index of refraction (*n*) and the speed of radar propagation (*c*) are plotted against altitude. Ray (*a*) is just able to skim over the top of the trapping layer, but ray (*b*) is bent by this layer so that no energy penetrates the region of the radar hole.

If the radar transmitter is above the trapping layer the radar hole is much larger than if the transmitter is below the layer, so that if the transmitter is brought far enough below the trapping layer beam splitting is lessened and the radar hole will disappear. Consequently, this problem is not serious for surface-based radars. However, for radars operating from aircraft it is necessary to determine whether trapping layers exist, and if so, to act accordingly.

In addition to the increased attentuation produced by rain mentioned in section 811, the very high frequency (VHF) radio waves (between about 50 MHz and 200 MHz) are susceptible to the effects of refraction. Although the speed with which electromagnetic radiation travels through the atmosphere is related to both temperature and water-vapor content, temperature has the major effect in the lower troposphere. Since an increasing temperature is associated with a higher speed of transmission for radio waves, in a region of a temperature inversion the radio waves tend to be bent backward toward the earth's surface, whereas under normal lapse rate conditions, the radio waves are bent away from the earth's surface.

One example of this is the case of a passage of a warm front (see Chapter 16) shown in Figure 8-5. It may be seen that within the region of temperature inversion the signal strength increases markedly as the radio energy is bent down toward the reception area. Similar effects will be produced by any meteorological condition which tends to produce a temperature inversion, such as cooling by radiation of the lower layers during the early morning hours.

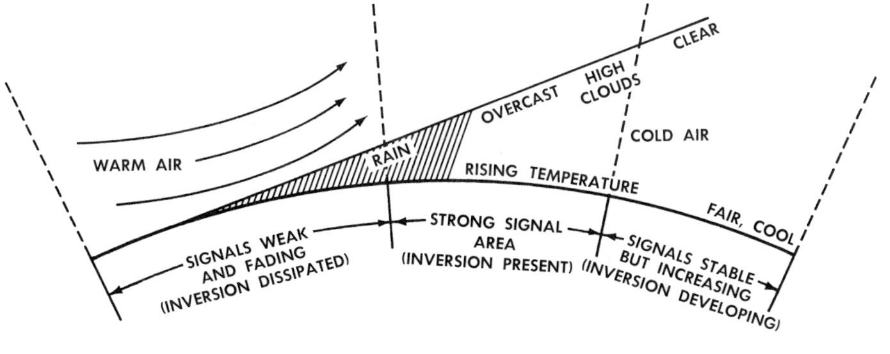

Figure 8-5 Effect of a Warm Front on Signal Strength

814 Sound. In contrast to electromagnetic energy, sound is a form of energy which is transmitted by a variation in pressure. It is a longitudinal wave which involves the motion of medium particles in the direction of energy transfer as the wave goes by. As the particles of the medium move back and forth, regions of rarefaction (low pressure) and compression (high pressure) are produced. The amplitude of these pressure swings is very, very small for sounds within the normal range of hearing. In the atmosphere, for example, the pressure variation associated with the smallest sound which can barely be heard by an average human ear is about 2×10^{-4} dynes per square centimeter (2×10^{-10} atmospheres) in amplitude. On the other hand, the variation in pressure produced by the loudest sound that can be heard without experiencing pain has a magnitude of 0.0002 normal atmospheric pressure.

The intensity of sound waves is proportional to the square of this pressure variation and is usually measured in watts per square centimeter. The equivalent intensity for the threshold of hearing indicated above is 10^{-16} watts per square centimeter; for the threshold of pain it is 10^{-4} watts per square centimeter.

When sound passes through air, the intensities and pressures which are involved are quite a bit different than those if the medium is water. This results from the fact that both the density and compressibility of the two fluids are quite different. For example, if the same acoustic pressure is present, the intensity of a sound in air is about 3,600 times the intensity that would be involved in water. On the other hand, if the same acoustic intensity is present the acoustic pressure in water is about 60 times that of the sound pressure in air.

815 The Decibel. Since the pressure and intensity magnitudes involved in the study of sound vary over such a tremendous range, it is desirable to develop a unit of measurement which allows for this without the necessity of large exponents. Such a unit is the decibel which is defined as follows:

$$\text{Sound level in decibels } (db) = 10 \log \left(\frac{I}{I_0}\right)$$

$$db = 10 \log \left(\frac{p}{p_0}\right)^2$$

$$db = 20 \log \left(\frac{p}{p_0}\right) \qquad (10)$$

where I_0 is the reference intensity and
p_0 is a reference pressure.

Since the logarithm of 2 is about 0.3, a doubling of the intensity will result in a change in sound level of 3 decibels. It also turns out that if the approximate threshold of hearing for the average human ear (2×10^{-4} dynes per square centimeter at 1000 Hz) is utilized as a reference pressure, an increase in sound level of 1 decibel is about the minimum change that can be distinguished by the average observer.

In air the threshold of hearing, 2×10^{-4} dynes/cm^2, is usually used as a reference level. However, in water, it is more convenient to use a reference pressure of 1 microbar (1 dyne per square centimeter). To convert from decibels using 1 microbar as a reference to decibels using 2×10^{-4} dyne per square centimeter as a reference, simply add 74

Table 8-3 Sound Levels in Air and Water

AIR (re 0.0002 dynes/cm^2)	WATER (re 1 microbar)
80 feet from take-off of jet aircraft (135 db)	100 yards from underwater dynamite explosion (61 db)
threshold of human aural discomfort at 1,000 hz (120 db)	nearby 25-HP outboard motor (46 db)
15 feet from single-propeller aircraft (100 db)	very rough sea in conjunction with heavy rain (26 db)
20 feet from full symphony orchestra (90 db)	chorus of several hundred marine catfish (16 db)
very noisy business office (80 db)	noise of ships in a busy harbor (6 db)
loud home Hi-Fi set (74 db)	100 yds from thousands of snapping shrimp (0 db)
average home radio (64 db)	calm sea (−10 db)
quiet residence (32 db)	threshold of hearing for goldfish (−42 db)
threshold of human hearing at 1,000 hz (0 db)	(−74 db)

decibels. For example, 0 db with 1 microbar as a reference is equivalent to 74 db with 2×10^{-4} dynes/cm^2 as a reference. To acquaint the reader with decibel levels associated with various sounds, both air-borne and water-borne, Table 8-3 is given.

Note that the two decibel scales are used side by side and the sounds in air and sounds in water may therefore be specified with either scale, if the reference level is indicated.

816 Sound Absorption. Because absorption is less in water than it is in air and sound speed is greater in water than it is in air, water is a much better medium for the propagation of sound. The absorption in both water and air is a function of the frequency and it increases as the frequency increases, so that higher frequencies are absorbed to a greater extent in both water and air. At a frequency of 1 kHz, the characteristic length in air is about 10 kilometers, whereas in water it is about 400 kilometers; at 10 kHz a typical atmospheric £ is about 100 meters, whereas in sea water it is about 4 kilometers. Throughout the audio spectrum—from 20 hertz up to 20,000 hertz—the absorption losses in the ocean are markedly less than those in the atmosphere.

817 Energy Loss Comparison. It might be desirable at this point to compare the absorption of light, radio waves, and sound waves in both the atmosphere and the hydrosphere. This comparison is tabulated conveniently in Table 8-4.

Table 8-4 Energy Losses in Atmosphere and Hydrosphere

ENERGY TYPE	ATMOSPHERIC £	HYDROSPHERIC £
Radio at $f = 18$ kHz	10,000 kilometers	1.69 meters
Radar at $f = 3{,}000$ MHz	600 kilometers*	4.13 millimeters
Blue-green light	45 kilometers	100 meters
Sound at $f = 1$ kHz	10 kilometers	400 kilometers
Sound at $f = 10$ kHz	100 meters	4 kilometers

Note that of all forms of energy, radio waves are absorbed least in the atmosphere, whereas in the hydrosphere sound waves are least attenuated. It now should be apparent why radio is used for communication in the atmosphere and sound for ranging and location within the ocean.

* Actually $£_a$.

818 Sound Scattering. Another loss, fairly common to all wave energies, is scattering. Sound energy is scattered in both the atmosphere and the hydrosphere by any objects having markedly different sonic characteristics than the medium. This is especially true in the ocean for proper size bubbles. Gas bubbles in the ocean may scatter sound up to 1,000 times as much as particulate matter of the same size, due to resonance effects.

An important practical application of sound scattering in the ocean is the presence of the *deep scattering layer* (DSL). Although composed of a large organic community ranging from small zooplankton to large squid (section 1406), the sound scattering is produced by small gas bubbles associated with particular members of the community.

819 Spreading Losses. Generally speaking, in both the atmosphere and the hydrosphere, the greatest loss of sound energy is due to the spreading of energy as it travels from the source. This may take two basic forms: *cylindrical spreading* where the intensity at a distance is inversely proportional to the distance, or *spherical spreading* where the intensity at a distance is inversely proportional to the square of the distance. Another way of expressing this is to say that with cylindrical spreading, losses are 3 decibels per distance doubled, whereas with spherical spreading, losses are 6 decibels per distance doubled. It will be seen very shortly (section 822) that the most marked increase in sound range both in the atmosphere and in the hydrosphere may be accomplished by limiting this spreading loss.

820 Sound Speed. In the atmosphere the speed of sound is directly related to the temperature. The expression for dry air sound speed is as follows:

$$c = 20.08\sqrt{T} \qquad (11)$$

where

T is the air temperature in degrees Kelvin, and

c is the sound speed in meters per second.

With the addition of moisture to the air, sound speed will change slightly; the change is so slight that it can usually be neglected.

In the ocean, sound speed is also related to the temperature, and to a lesser extent the salinity and the pressure. An increase in any of these three parameters will produce an increase in sound speed. A temperature increase of 1 degree Celsius will produce an increase in sound speed of between 3 and 4 meters per second. A salinity increase of 1 part per thousand will produce an increase in sound speed of about 1 meter per second, while a depth increase of 55 meters will produce an increase in sound speed of 1 meter per second. These changes are superimposed on an average oceanic sound speed around 1,450 meters per second, whereas in air the average sound speed is about 330 meters per second.

In both the atmosphere and the hydrosphere, for all practical purposes, sound speed does not vary with frequency. Consequently, it will be assumed that any sound speed determination will be applicable for all frequencies, no matter at what frequency the determination is made.

Salinity, temperature, and pressure all vary in the ocean, so that a knowledge of all three appears to be necessary to determine the sound speed. However, sound speed is affected more by temperature change than salinity, and in addition, the total variation in temperature to be expected in the ocean is much greater than any variation expected in salinity. Thus, to a very good first approximation a measurement of temperature variation is sufficient to determine the variation of sound speed accurately enough for naval operational purposes. Interestingly enough, in both the atmosphere and the hydrosphere the variation of temperature is of prime import as far as sound speed is concerned; an increase in temperature will produce an increase in sound speed.

821 Sound Ranges. With the normal lapse rate condition in the atmosphere, sound waves will be bent away from the surface of the earth making long-distance sound ranges unusual, to say the least. However, long sound ranges occasionally exist when temperature inversions are found. An example of this is associated with the Krakatau Volcano in Sunda Strait, between Sumatra and Java, which erupted on 27 August 1883. Apparently the sound generated by this volcanic activity was so great that, after having been refracted back to earth by the temperature inversion in the stratosphere, it still had sufficient energy left to be heard as far away as 4,775 kilometers.

822 Ducts and Channels. In the hydrosphere, the temperature structure is often ideal for long-range transmission in the surface layers. The normal type of temperature variation with depth is such that the temperature is usually fairly constant in the mixed layer, under which is a thermocline, and then a deep layer of relatively isothermal cold water. With this situation, a slight increase in speed with depth is observed in the upper mixed layer due mainly to the effect of increasing pressure. The sound will be refracted toward the surface since the sound speed is greater at greater depths. This results in what is called a *surface duct*, in that any source within the mixed layer has its sound energy more or less trapped by refraction below and reflection at the water-air interface above.

Some of the sound energy however, will leave the source at such an angle that it will be refracted down as it passes through the thermocline since the sound speed is greater above. Refraction continues in this manner until the bottom of the thermocline is reached. Here again the sound-speed profile changes, exhibiting an increase in speed with increasing depth. This last change occurs as a result of the increasing pressure in the deeper isothermal layer causing the sound speed to increase with a resulting upward refraction (Figure 8-6).

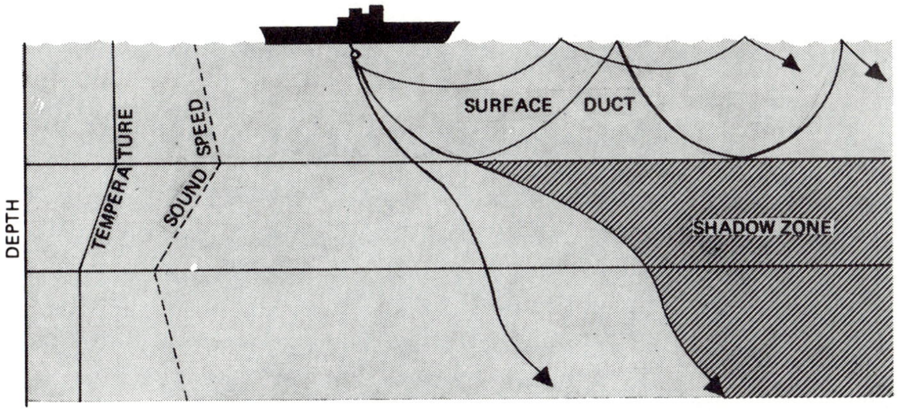

Figure 8-6 Typical Sound Rays for an Upper Isothermal Layer Showing the Surface Duct and Shadow Zone. (Note the similarity between the radar hole in Figure 8-4 and the shadow zone.)

Note that at the bottom of the thermocline a sound speed minimum may be found. This sound speed minimum serves as an axis of another sound layer, in this case called the SOFAR (Sound Fixing And Ranging) Channel. A sound source placed at the axis depth (the sound speed minimum), will produce sound which is refracted downward in the upper portion and upward in the lower portion, effectively trapping the sound within this channel.

In other words, the sound-spreading loss in both the duct and the channel is diminished. The sound no longer can spread in a vertical direction, but is only allowed to spread in a horizontal direction, making the range intensity inversely related to the distance, rather than the distance squared. Use of the SOFAR Channel in the deep ocean has resulted in small explosions being heard for distances of 12,000 miles.

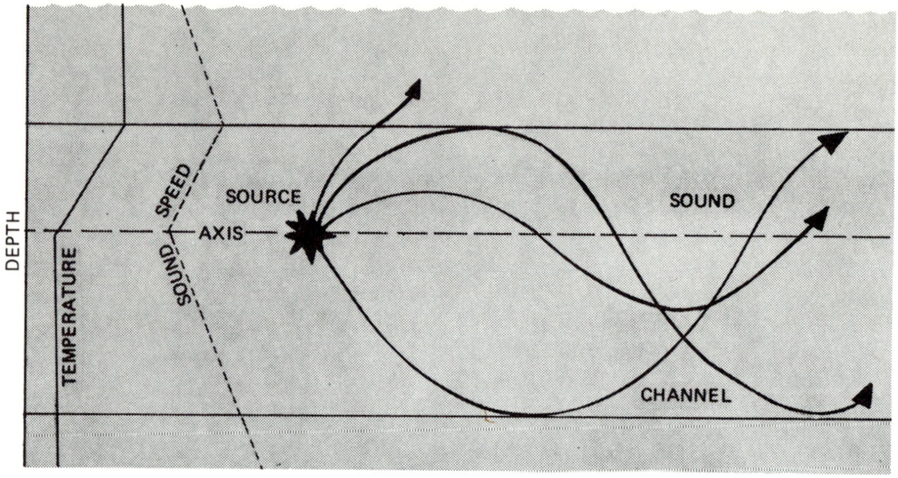

Figure 8-7 The Production of a Sound Channel By a Sound Velocity Minimum

One interesting and important aspect of the refraction of sound in the hydrosphere is the existence of surface sound ducts over large portions of the ocean. The sound is either channeled within this duct or else is bent down toward the bottom, leaving a region just below the sound duct wherein no sound penetrates. Just as in the case of the radar hole, no energy will penetrate this region so that a target may lurk here and not be detected. In the case of sonar, the region of no energy penetration is called the *shadow zone*. Unless the ASW unit is aware of the zone's location, a submarine may very easily hide in it without being detected. A large portion of ASW operations is designed to utilize sonar in such a manner that this shadow zone is covered with sound energy. This, of course, may be accomplished simply by moving the source around, in the vertical direction by the use of Variable Depth Sonar (VDS) and in the horizontal by moving the ship.

823 Sonar Transducers. Just as radar is utilized in the atmosphere for location and identification of targets, sonar or sound energy is utilized in the hydrosphere since this is the form of wave energy which has the smallest attenuation there. A sonar device is very simply a sound transmitter which sends out sound energy. If there is a target in the area, some sound energy reflects from the target and returns to the source. It is received by a hydrophone and converted into electrical energy which is then processed and presented in various forms (acoustic or video, for example).

The device which sends sound energy into the water (an underwater loud speaker) and receives sound energy from the water (an underwater microphone) is called a *transducer*, since it converts sound energy into electrical energy or vice versa. Two basic types of sonar transducers are available. One uses either the *electro-strictive effect* or the *piezo-electric effect*, where a *crystal* changes its physical dimensions when excited by an electrical field. This is the same type of device utilized in many phonograph pickups. The other type utilizes the phenomenon of *magnetostriction* wherein *nickel alloy rods* will change their length when excited by magnetic fields. Since a sonar transducer characteristically is involved with large amounts of acoustic power, synthetic crystals such as barium titinate are most suitable for these units. Transducers may be either hull-mounted or on the end of cable to allow for lowering into the hydrosphere to cope with varying temperature structures. These may be lowered from surface vessels (VDS), or helicopters (Dipping Sonar).

Figure 8-8 A Variable-Depth Sonar Transducer

824 Ambient Noise. In addition to the sounds that man puts into the water, whether they be from ships or from the sonar devices, there are natural sounds already present. These may take the form of sounds produced by waves, objects falling into the water, rain on the surface of the water, breaking ice, etc., or they may be produced by animals living in the water.

In any event, the sea is not silent. There is always a background of noise beneath the surface, and sometimes this background gets to be quite high as indicated by the decibel values given in Table 8-3.

Additional Reading

Albers, V. M., *Underwater Acoustics Handbook*, Pennsylvania State University Press, 1960.

Hill, M. N. (Ed.), *The Sea*, Vols. I & III, John Wiley & Sons, 1962.

Horton, J. W., *Fundamentals of Sonar*, 2nd ed., U. S. Naval Institute, 1959.

Humphreys, W. J., *Physics of the Air*, Dover Publications, Inc., 1964.

Jerlov, N. G., *Optical Oceanography*, Elsevier Publishing Company, 1968.

Middleton, W. E. K., *Vision through the Atmosphere*, University of Toronto Press, 1958.

Neuberger, H., *Introduction to Physical Meteorology*, University of Pennsylvania, 1957.

Neumann, G. and Pierson, W. J., *Principles of Physical Oceanography*, Prentice-Hall, Inc., 1966.

Officer, C. B., *Introduction to the Theory of Sound Transmission*, McGraw-Hill, 1958.

Urick, R. J., *Principles of Underwater Sound for Engineers*, McGraw-Hill, 1967.

Williams, J. *Oceanography*, Little, Brown & Co., 1962.

——, *Optical Properties of the Sea*, U. S. Naval Institute, 1970.

Introduction to Sonar Technology, Bureau of Ships, NAVSHIPS 0967-129-3010, Government Printing Office, Washington, D. C.

Meteorological Aspects of Radio-Radar Propagation, NAVWEPS 50-IP-550, Government Printing Office, Washington, D. C.

The Radio Amateur's Handbook, American Radio Relay League.

CHAPTER NINE

Winds and Currents

901 The Energy Source. Chapter 5 describes how solar energy caused heat to be added to both the atmosphere and the hydrosphere. This addition of heat affected sea and air temperatures, and produced large amounts of evaporation from the hydrosphere. Uneven heating resulted from both variations in absorption and reflection properties, and differing specific heat capacities of various materials over the earth.

Uneven heating tends to produce density differences in both the atmosphere and the hydrosphere. These density differences are the energy sources necessary to run the environmental motion machine, producing winds in the atmosphere which create currents in the hydrosphere. These wind-driven currents are in addition to the currents produced by density differences in the oceans themselves.

This chapter describes how density differences, produced by uneven heating, result in forces which cause the fluids of the atmosphere and the hydrosphere to move. This will be preceded by the development of a few basic mathematical tools which will be of considerable help.

Swell breaks over San Francisco Bar

902 Parameter Representation. The environment is very often represented in pictorial form by plotting data in the form of *isopleths*. An isopleth (Greek, *isos* + *plethes* – equal quantity) is a general term used to define a line of constant value for any parameter. An *isotherm* is an isopleth joining points of equal temperature; an *isohaline*—equal salinity; *isopycnal*—equal density; and an *isobar*—a line of equal pressure.

To observe how these lines are plotted, first look at a typical topographical feature, in this case a hill. The sketch in Figure 9-1 shows a hill which is somewhat over 200 feet above sea level. If the hill is cut by two level planes, one at 200 feet above sea level and the other at 100 feet above sea level, the surfaces produced by these planes cutting through the hill may be represented by the same topographical map. Note that there are three nesting curves for this particular hill, one labeled 200 feet, one 100 feet, and the other at sea level. The curve labeled 200 feet represents the intersection of a plane 200 feet above sea level as it cuts through the hill; while that labeled 100 feet represents the intersection of the hill and a plane 100 feet above sea level.

A valley or a crater may be represented in a similar manner. Since variations in height in this case are measured from the bottom of the

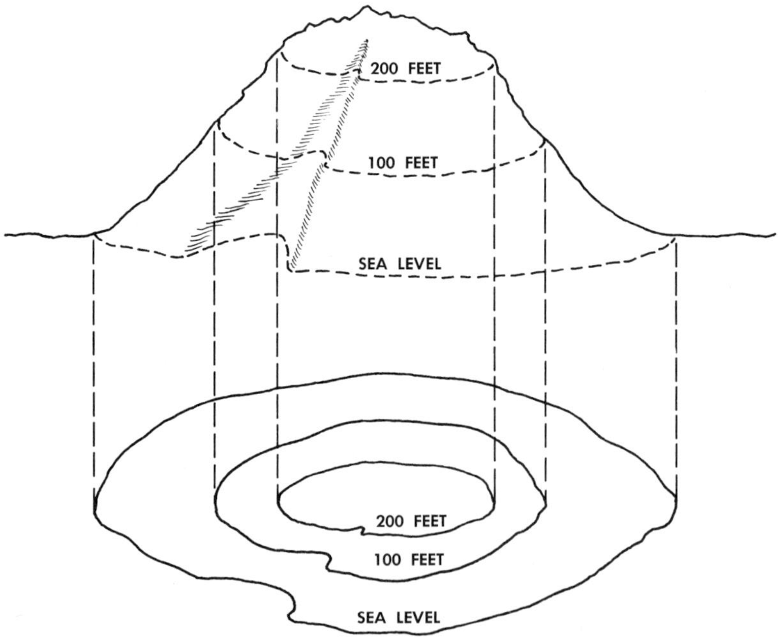

Figure 9-1 Topographic Map of a Hill

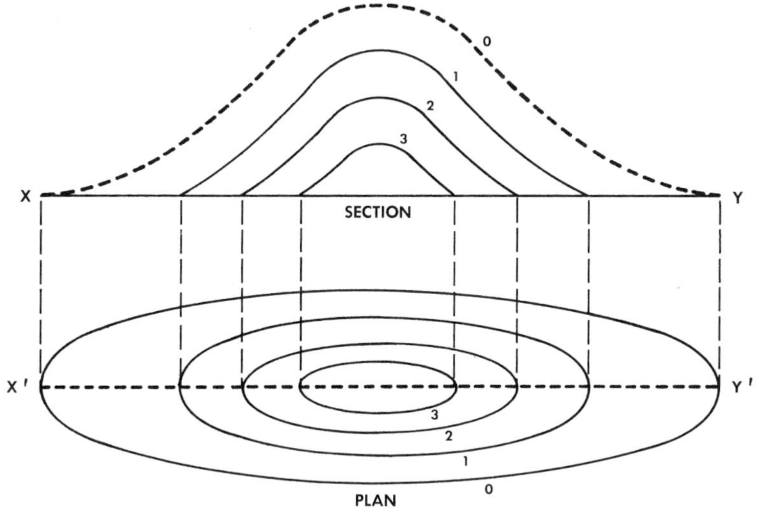

Figure 9-2 Section and Plan of Isobars Produced by a Mound of Fluid

crater, the number associated with the crater interior is smaller than that associated with the exterior. A crater, therefore, represents a low in the surface, whereas a hill represents a high.

903 Plotting Pressure Fields. Pressure may also be plotted using topographical type charts. Imagine a mound of fluid. This mound may be represented in a like manner to the hill discussed previously. However, the pressures existing at some level within the mound may be represented by cutting a reference plane through that level. If all points having zero pressure are joined, a zero pressure curve called the *zero isobar* may be drawn. The one-unit isobar represents the line along which the pressure is equal to one and is also the intersection of the reference plane and the one-unit pressure surface. Note that a chart very similar to a topograph chart results, but in this case, pressure patterns which exist are plotted on a level surface.

It may be seen that, anywhere along the line labeled 2, the fluid above this line has a column height which is greater than that for the fluid above the line labeled 1. In general, pressure is given by the following formula:

$$p = \bar{\rho}gh \tag{1}$$

as discussed in section 408.

It may be seen that with a greater column height an increase in pressure would be expected. For the pressure mound sketched in Figure 9-2, the height of the column of fluid above the reference plane gets greater toward the center so that the pressure also increases toward the center. This is a high-pressure area.

A low-pressure area may be represented in a similar manner, as is done in Figure 9-3. Since the fluid has mass, the pressure increases as the height of the fluid column is increased. Consequently the top pressure line is labeled 0, the one underneath is 1, the one below it is 2, and so on. In this illustration the same series of nesting curves result;

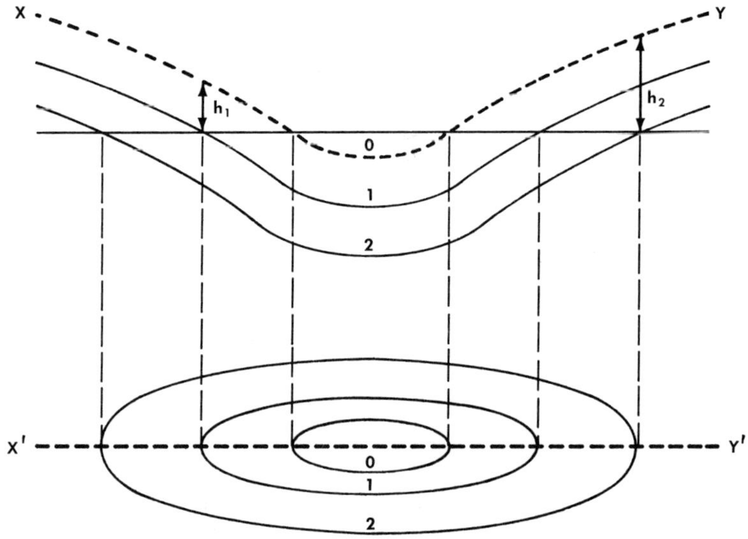

Figure 9-3 Section and Plan of Isobars Produced by a Fluid Depression

however, the isobar in the center represents the intersection of the zero pressure line and the level surface rather than a higher value. In this particular case the topography represented is that of a low-pressure area or a depression. Note that, as before, the greater the height of the column above the reference level, the greater the resulting pressure.

904 Vector and Scalar Quantities. In many physical phenomena, it is desirable to represent quantities which have direction in addition to having a magnitude. In order to do this vectors may be used. A vector is represented by an arrow; the direction in which the arrow points indicates the direction of the parameter, and the length of the arrow itself is indicative of the parameter's magnitude. Suppose, for example, the vector sketched is one and a half inches long and points to the north. It may represent a velocity of 15 knots directed toward

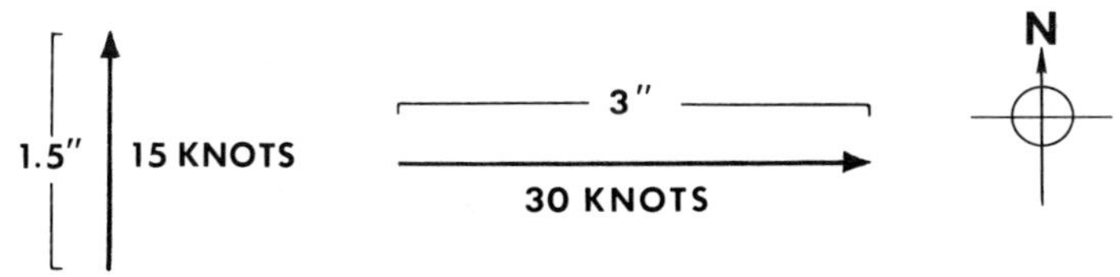

Figure 9-4 Two Vectors

the north. Next to it is another vector which is 3 inches long and is pointed toward the east. Using the same scale, this vector would represent a velocity of 30 knots directed toward the east. Vectors may be used for any type of quantity which has *magnitude and direction.* Examples are velocity, acceleration, and force. Examples of *non-vector* or *scalar quantities* are mass, color, frequency, and speed. Note the difference between speed and velocity.

An interesting sidelight to the utilization of vectors is the convention which is employed when rotational motion is encountered. Due to a possible difference in the radius of curvature, it is difficult to draw circular vectors and have them be meaningful. By convention a motion which is rotational is represented by a straight-line vector using the right-hand rule. Sketched in Figure 9-5 is a disc, rotating counterclockwise as viewed from above and represented by a linear vector pointing along the axis of rotation, in this case, upward. The direction is obtained from the direction of the right thumb when the fingers of the right hand are curled in the direction of rotation. As before, the length of the vector is proportional to the magnitude of the quantity involved. If the rotation sketched were reversed, the right thumb would now point down, indicating a reversal in direction.

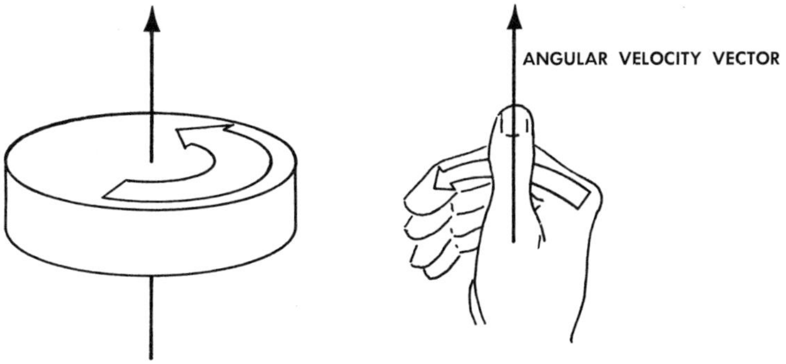

Figure 9-5 The Right-Hand Rule

905 Addition of Vectors. If one wants to go from *A* to *B* when *A* is southwest of *B*, it may be easier to use the existing sidewalks by going a block to the east and a block north. The resulting motion may be described by either the sum of two motions, one a block east and the other a block north, or by one direct action, 1.414 blocks to the northeast. Similarly, the true wind direction and speed may be determined at sea by adding the ship-motion vector and the relative wind vector.

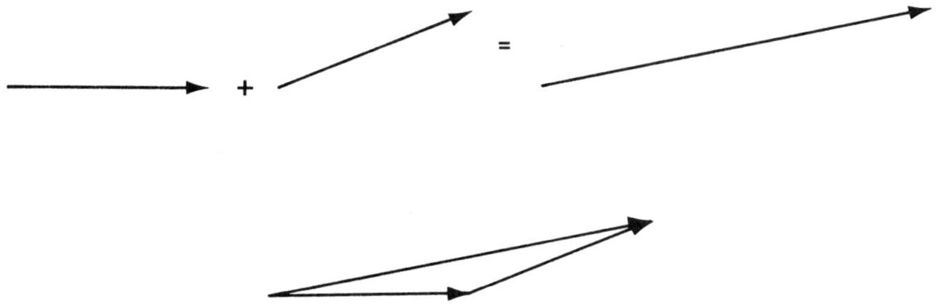

Figure 9-6 Adding Vectors

In other words, vectors may be added, and the manner in which vectors are added is by means of the following convention: if a vector *A* is to be added to a vector *B*, the tail of vector *B* is placed against the head of vector *A*, being careful to maintain the relative direction of both vectors. The sum of these two vectors is simply the vector starting at the tail of vector *A* and ending at the head of vector *B*, as indicated in Figure 9-6.

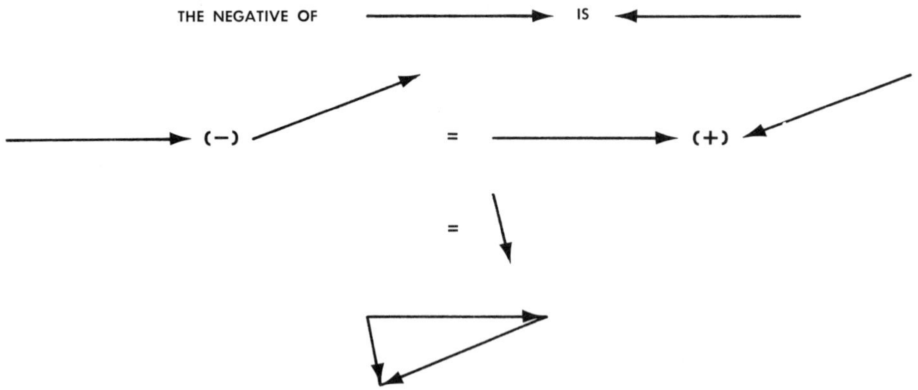

Figure 9-7 Subtracting Vectors

Vectors may be considered to have signs. If a vector is plus, a vector of equal magnitude but oppositely directed may be considered to be the negative equal of the former. If vector A is multiplied by -1, its direction is simply changed by 180°. This concept of the negative vector is illustrated in Figure 9-7. It is now a simple matter to subtract vectors. All that must be done is to add the negative of the second to the first, $A-B=A+(-B)$.

906 Vector Components. One last vector property of interest is that of components. Any vector may be broken down into components which may be thought of together as exactly representing the initial vector. Assume a vector A of 5 units in length directed at an angle of 37°* to the horizontal. This vector may be considered to be composed of a vector 4 units long in a horizontal direction plus one 3 units long in a vertical direction. In other words, one set of components for a vector 5 units long consists of one vector 3 units long and one 4 units long at right angles to each other. Any vector may be represented by any two vectors which when added vectorially result in the initial vector. Some examples of replacing a vector by various sets of components are sketched in Figure 9-8.

It is often convenient in the environmental sciences to replace various vectors by components which are perpendicular and directed along the coordinate axes chosen. The three axes are usually chosen so that the X axis has an east-west orientation, the Y axis a north-south orientation,

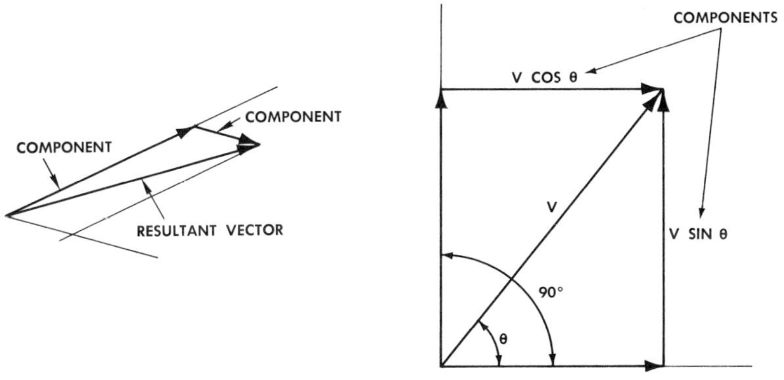

Figure 9-8 Two Examples of Vector Components

* This is an approximation since the sine of 36°52′ is equal to 0.59995.

and the Z axis is oriented vertically along a radius of the earth. Perpendicular components, directed along these three axes will always be used in this text.

907 Pressure Gradient Force. Of the more common forces normally produced in the environmental fluids, the most prevalent is the pressure gradient force (*P*). This force has both horizontal and vertical components, but the former is the important one where winds and currents are concerned. The horizontal component of the pressure gradient force, a force produced by a difference in pressure along a horizontal distance, may be caused by mechanical action such as water piled up by a wind. It may also result from differences in density within the fluid, since the pressure of a fluid column is proportional to the height and the average density of the column. As a matter of fact, whenever there is a pressure gradient in a fluid, there is also a density gradient associated with it. These two are coexistent phenomena.

The pressure gradient force is a function of the slope of the pressure surface, and is directed from high to low pressure. If the sides of the mound of fluid are very steep, a very-high-pressure gradient force results, whereas if the slope of the mound is very gentle, the pressure gradient force is relatively small. In the atmosphere another way of describing this is to indicate that with closely spaced isobars, the pressure gradient force is high, and with widely spaced isobars the pressure gradient force is low. This is simply another way of expressing the relative steepness as indicated before.

908 Coriolis Force. Once the fluid is set in motion by one means or another an additional force enters into the picture; that is *coriolis force*. This force is produced by the effect of the rotating earth, and

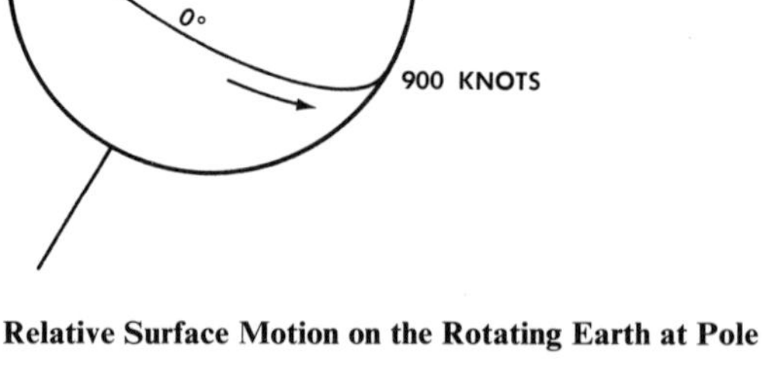

Figure 9-9 Relative Surface Motion on the Rotating Earth at Pole and Equator

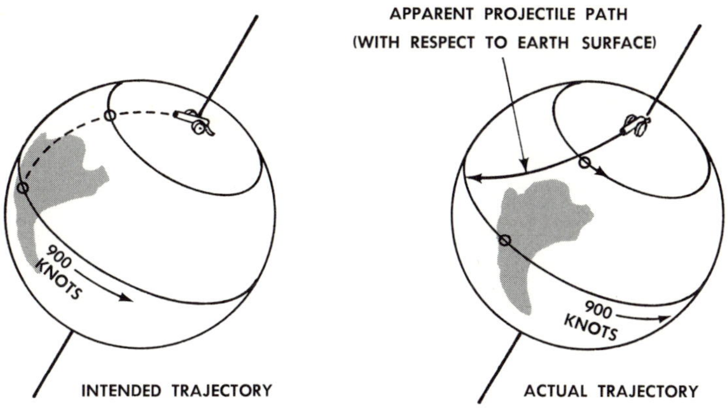

Figure 9-10 Projectile Deflection

since it is beyond the physical experience of most individuals, it will be discussed in some detail. At the North Pole the linear speed of the earth's surface produced by the earth's rotation is zero (Figure 9-9), since an observer at this point is standing on the top of the axis. However, as one leaves the North Pole, the linear speed of the earth's surface increases until at the equator the linear speed is about 466 meters per second (greater than 900 knots).

Now imagine a target on the equator and a gun crew at the North Pole (Figure 9-10). Somehow or other the gun crew can see the target and can aim their gun accordingly. The object of the exercise is to hit the target. The gun crew fires their gun, but since the target is moving at a different linear speed than the gun, they will miss the target. As a matter of fact, if the projectile had a speed of 1 kilometer per second (1,940 knots), by the time it reached the equator it would miss the target by something like 4,600 kilometers, a little more than 1/8th of the equatorial circumference. Now it is obvious why the gun crew missed their target; the target was moving and they did not allow for this motion. But to the gun crew it would appear to them that there was a force acting on their projectile, tending to deflect it to the right. As far as they could see, the target was not moving with respect to them since both the gun and the target were on the same surface rotating together. It may similarly be shown that a projectile fired from the equator toward the North Pole would again undergo a deflection to the right.

If the reader desires to try this same experiment on a smaller scale, he might attempt to play catch on a carnival merry-go-round while it is in motion. He would find that it would be extremely difficult to throw the ball so that it could be easily caught, for the same basic reason that the gun crew had difficulty. The linear speed of two different points on a rotating surface will be different, simply due to the rotation. This apparent deflection of a moving body, due to the rotation of the earth, is called the coriolis deflection, and the force which produces the deflection is called the coriolis force.

In Figure 9-11 it may be seen that the vector representing the angular velocity of the earth (Ω) is pointing up through the North Pole in the direction indicated according to the right-hand rule convention discussed previously. This vector may be moved at will on the earth's

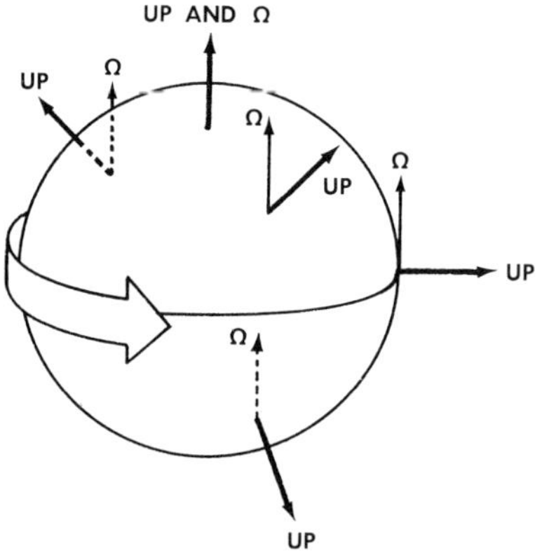

Figure 9-11 The Earth's Angular Velocity Vector at Various Latitudes

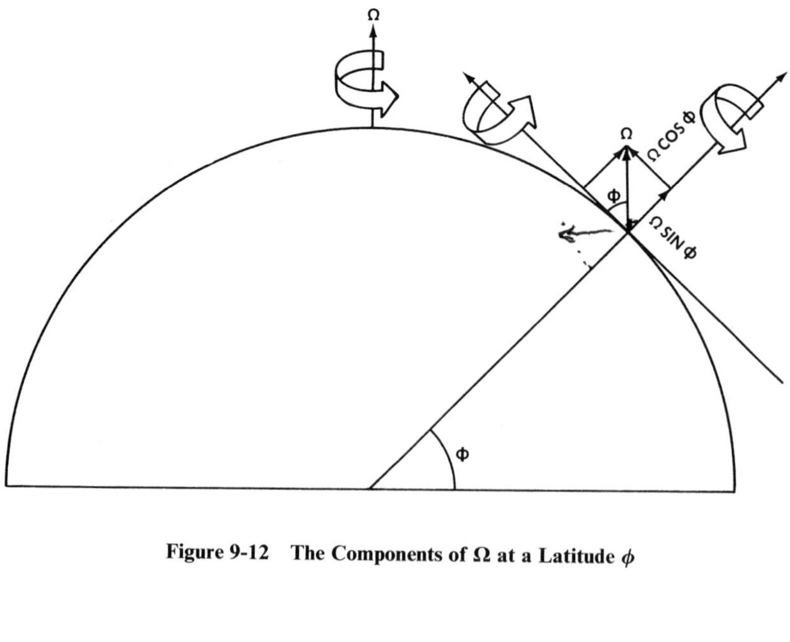

Figure 9-12 The Components of Ω at a Latitude ϕ

surface as long as its orientation is maintained. At some intermediate latitude it is still pointing in the same direction with respect to the fixed stars, but with respect to the earth's surface its direction has changed. This vector may be resolved into two components, one perpendicular to the surface of the earth at this point and the other tangent to the surface of the earth at this point (Figure 9-12). Since these two components are equivalent to the resultant at this geographical point, the earth's rotation may be considered equivalent to two separate rotations—one through an axis perpendicular to the earth's surface and another through an axis parallel to the surface at this point.

Since the vector which is perpendicular to the surface of the earth at this point is in the same relative position as the earth's rotational vector at the North Pole, the point of interest may be considered as acting like a North Pole about which the earth is rotating with an angular velocity of $\Omega \sin \phi$. As may be seen from the sketch, ϕ is the latitude and Ω is the angular speed of the earth's rotation, i.e. 2π radians per 24 hours. In other words, any motion away from this "pole" results in relative motion due to the rotation of the earth identical in concept to that experienced when moving away from the real North Pole.

The similarity may be shown quite easily by observing the operation of a Foucault pendulum. The Foucault pendulum is a very long pendulum hanging from a point which consists of a frictionless swivel. This is sketched in Figure 9-13. If the Foucault pendulum were set swing-

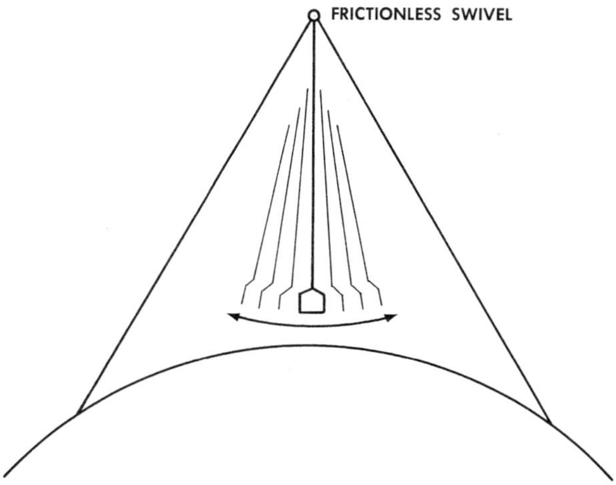

Figure 9-13 A Foucault Pendulum

ing at the North Pole, the earth would rotate under it as the pendulum swung back and forth. An observer would see an apparent motion by the plane of the pendulum swing having a period of 24 hours, and a rotational speed of Ω ($2\pi/24$ radians per hour, since $\Omega = 2\pi/T$). If a Foucault pendulum is placed at some other latitude, there is the same type of rotation by the plane of the pendulum swing; however, the rate at which the plane of swing rotates turns out to be $\Omega \sin \phi$. Note that this expression for the rotational speed of the Foucault pendulum is identical with the vertical component of the earth's angular speed which was derived previously by simple resolution of vectors.

The coriolis force is a function of the object speed and the vertical component of the earth's rotational speed at the point of interest.

$$C = 2\Omega v \sin \phi \qquad (2)$$

where: C is the coriolis force per unit mass,
Ω is the angular speed of the earth's rotation,
v is the speed of the moving body, acted upon by the coriolis force and,
ϕ is the latitude at which the force is measured.

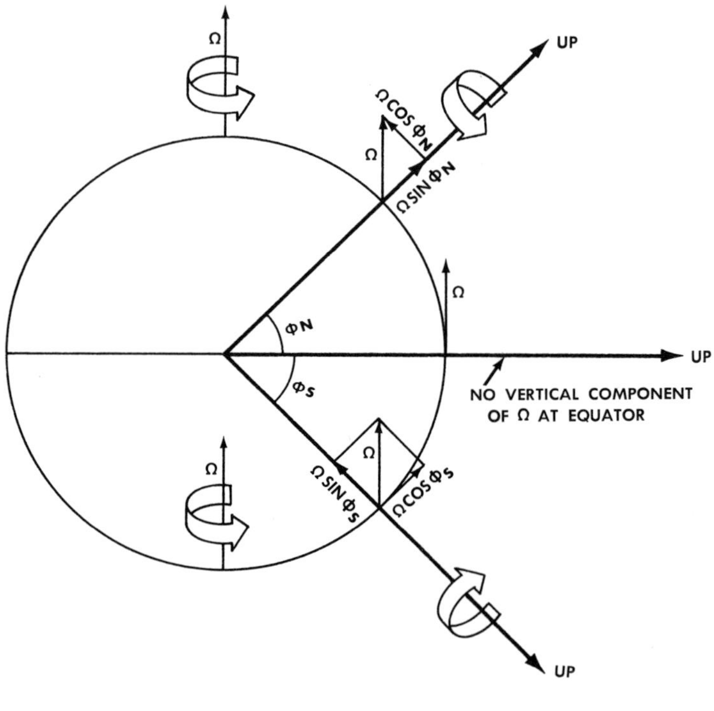

Figure 9-14 The Earth's Angular Velocity in Both Hemispheres

Note that at a latitude of 0°, the coriolis force vanishes. This is to be expected, since at the equator the vertical component of the earth's rotational vector vanishes, as may be seen in Figure 9-14. In addition, it may be seen that as one passes the equator and moves into the Southern Hemisphere, the vertical component of the earth's angular velocity changes its direction. In the Northern Hemisphere the vertical component points away from the center of the earth (up), whereas in the Southern Hemisphere the vertical component points towards the center of the earth (down). In other words, in the Southern Hemisphere the apparent rotation is opposite to that in the Northern Hemisphere. This has the effect of causing the coriolis force to always be to the right of the motion in the Northern Hemisphere and to the left of the motion in the Southern Hemisphere.

909 Centrifugal Force. When the fluid motion takes a curved path, it often becomes necessary to consider the effect of centrifugal force (S), the force produced by a change in direction of the curved path. The centrifugal force per unit mass is given by the following expression:

$$S = \frac{v^2}{r} \tag{3}$$

where: v is the speed of the moving body acted upon by centrifugal force and,
r is the radius of curvature of the motion.

This force is always directed *perpendicular* to the direction of motion and away from the center of curvature. The centrifugal force is usually quite important in relatively small, well-developed storms where high wind speeds and small radii of curvature are found. Examples of this are hurricanes and other tropical storms.

910 Wind Stress. A force very common to the production of motion in the hydrosphere is *wind stress force* (W), that is, the force of wind on the water surface. This, of course, is in the direction of the wind and has a magnitude related to the wind speed. Although the exact variation of wind stress with wind speed is somewhat in doubt, it will be enough to know that the greater the wind speed, the greater the wind stress force.

911 Friction. The last force to be considered is friction, which is important in both hydrospheric and atmospheric motion. *Frictional force* (F) is a function of both surface roughness and speed of flow, and is directed opposite to the motion at all times. It tends to produce a slower speed than if friction were not present.

912 Geostrophic Flow. There are many occasions in nature when a steady-state condition is found in the atmosphere or the hydrosphere. *Steady state* exists when there is only unaccelerated motion in the direction of flow; that is, the speed is a constant. With constant speed conditions there must be a balance of forces present, since a force greater in one direction than another would produce an acceleration.

One type of unaccelerated flow found in nature is called geostrophic flow. *Geostrophic flow* is motion in straight lines under conditions such that all forces except two are so small they may be neglected. These two are the pressure gradient force (P) and the coriolis force (C) which are in balance for geostrophic flow, i.e.

$$P = C \tag{4}$$

Suppose a pressure gradient force exists which tends to cause the fluid to flow downhill, toward the low pressure. The fluid will start out flowing downhill, but as soon as it starts to flow it will be acted upon by the coriolis force which is a function of the fluid speed. This first stage is sketched in Figure 9-15.

Since there is initially an inbalance of forces before this flow gets established, the tendency will be for the flow to change direction and be deflected toward the right (in the Northern Hemisphere). In the second stage there has been some deflection so that the fluid is veering to the right as it flows downhill. The pressure gradient force is still acting downhill, but now the coriolis force has changed direction, since it is

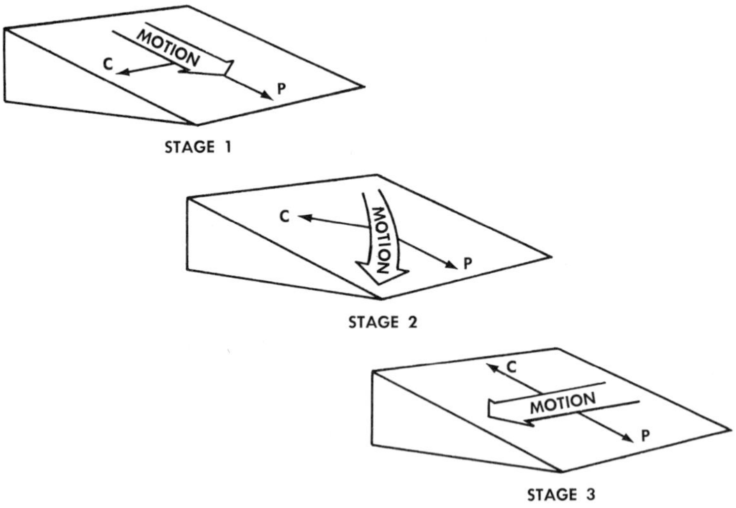

Figure 9-15 Stages in the Development of Geostrophic Flow

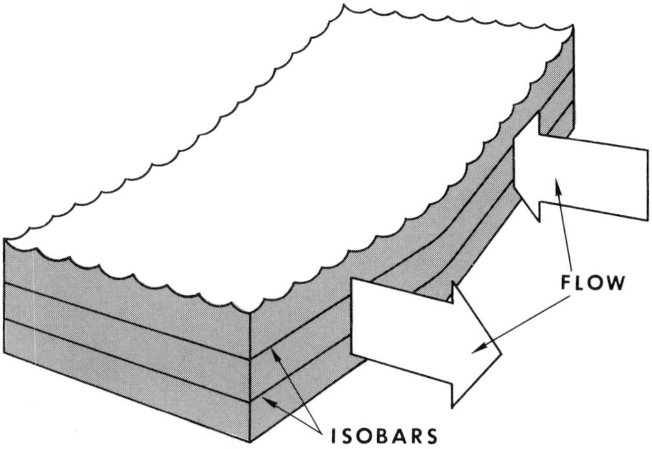

Figure 9-16 Direction of Flow Produced by Sloping Pressure Surfaces in Northern Hemisphere

always 90° to the right of the actual motion. Finally in the third stage, the coriolis force has pulled the direction of motion around until the flow is no longer downhill but is across the slope. The pressure gradient force acting to pull the fluid downhill and the coriolis force acting to the right of the motion (in the Northern Hemisphere) tending to pull it uphill are exactly equal, so there is a balance of forces. The deflection will never exceed 90° from the initial downhill direction since the pressure gradient force itself would oppose this action. Since these are the only two forces present, there is no acceleration and the speed is constant.

If one determines the pressure gradient force from the slope of the pressure surfaces, one may then calculate the resulting current or wind, if geostrophic flow is present, by simply equating the pressure gradient and coriolis forces. Note that in geostrophic flow the direction of flow is parallel to the isobars. The flow is not downhill but across the hill, and it is directed as indicated in Figure 9-16. Its speed will depend upon the magnitude of the slope; the greater the slope the greater the speed.

913 Gradient Flow. If geostrophic flow is assumed, but along a curved path rather than a straight-line path, then the centrifugal force must be reckoned with. If the pressure gradient force, the coriolis force, and the centrifugal force are the only three forces of importance, the resulting motion is called *gradient flow*. From the vector diagram below a balance of forces again exists; in this case,

$$P = C \pm S \tag{5}$$

It may be seen that geostrophic flow is a special case of gradient flow wherein the radius of curvature is infinite.

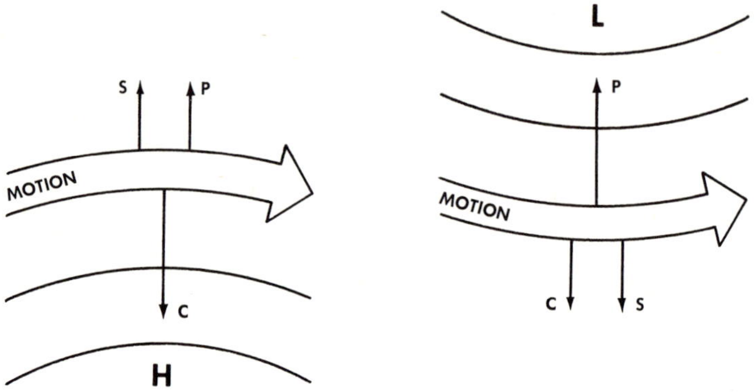

Figure 9-17 The Forces Involved in Gradient Flow in the Northern Hemisphere

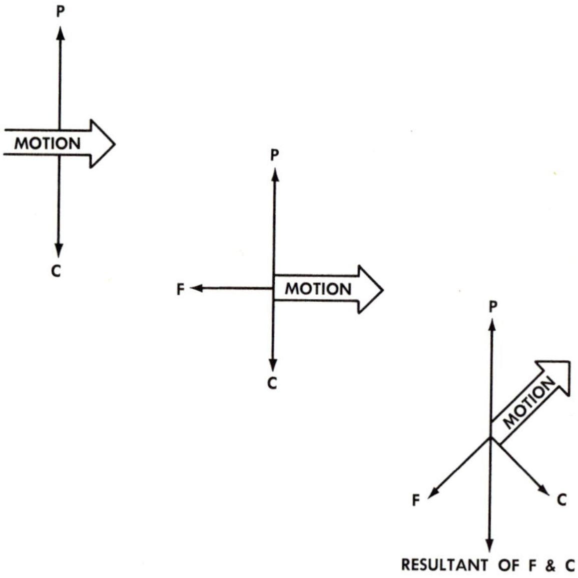

Figure 9-18 The Effect of Friction on Geostrophic Flow (S=0) in the Northern Hemisphere

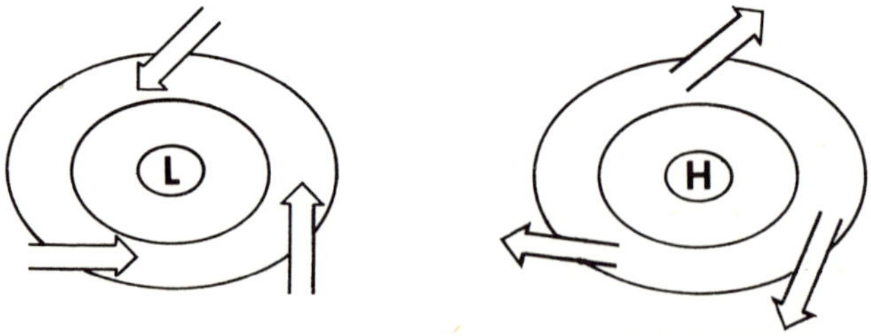

Figure 9-19 Frictional Flow with a Low and a High

Just as in the case of geostrophic flow, the direction of gradient flow is parallel to the isobars and is determined in exactly the same manner as before. This results in a clockwise circulation around a high-pressure area and a counterclockwise circulation around a low which is sometimes expressed by meteorologists as *Buys Ballots' law*. This law states that when facing the wind, low pressure will be to the right in the Northern Hemisphere and to the left in the Southern Hemisphere.

914 Frictional Flow. Suppose the force of friction (F) is now introduced. What effect does this have on the two types of flow discussed above? In general, friction will have a twofold effect. It will, as would be expected, tend to decrease the speed of flow, and, due to this decrease in speed, it will also tend to change the direction of flow.

In the vector diagram in Figure 9-18 is shown a case of geostrophic flow in the Northern Hemisphere. There is a balance of forces between the pressure gradient and the coriolis; they are equal and oppositely directed, with the resulting motion in a direction perpendicular to both. If the force of friction is now introduced, it will directly oppose the motion tending to retard the geostrophic flow. As the motion slows, the coriolis force will decrease because it is dependent upon the speed. The pressure gradient force does not change since it is related only to the slope of the pressure surfaces. With a change in speed and an associated change in coriolis force, there is an imbalance of forces as indicated in the sketch. Therefore, the direction of motion turns toward the left in the direction of the unbalanced force. As the motion turns toward the left, note that both the coriolis force, which is always to the right of the motion, and the frictional force, which always opposes the motion, also change in direction until a condition of balance is reached among the forces of friction, coriolis, and pressure gradient. There is now an unaccelerated flow of a slightly decreased magnitude in a direction to the left of the geostrophic flow. This is defined as *frictional flow*,

$$P = F + C \pm S \tag{6}$$

Thus friction has the effect of decreasing the magnitude of the speed and at the same time directing the flow across the isobars toward low pressure.

For this reason, a low-pressure area is a convergent region at sea level, since there is a tendency for fluid to flow in to the center. Conversely, a high-pressure area is a divergent region at sea level, since there is a tendency for fluid to flow outwards, as indicated in the sketch.

915 The Ekman Spiral. In 1893 F. Nansen set out on a rather remarkable expedition to the Arctic Ocean. He purposely allowed his

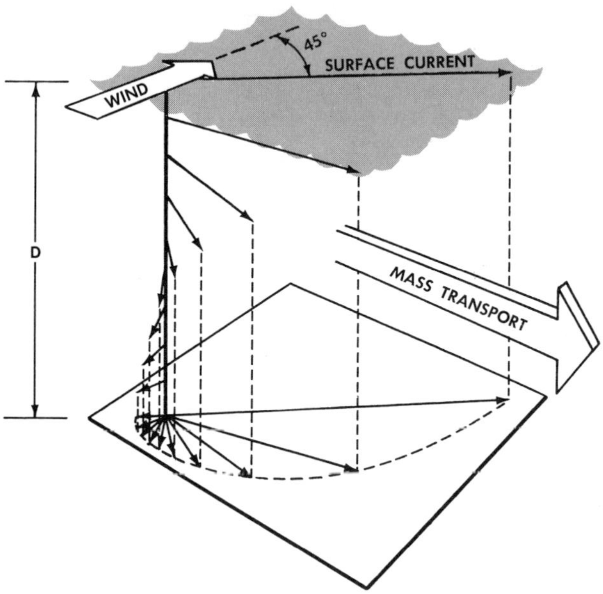

Figure 9-20 The Ekman Spiral in the Northern Hemisphere

specially designed vessel *Fram* to become frozen in the ice for three years to more accurately study the drift of the polar ice pack. One of his observations indicated that surface ice, when under the influence of wind, moved in a direction different to that of the wind. He communicated this information to V. W. Ekman who in 1902 published a theoretical explanation for this anomalous behavior. The effect of wind on the surface layer gave rise to the mathematical model called the *Ekman spiral.* To produce it Ekman set up a balance of forces between the coriolis force and the wind-stress force.

$$W = C \tag{7}$$

When this was done and the equations were solved, the result was a variation of current speed and direction as indicated in Figure 9-20.

At the surface the maximum current value, usually about one to two percent of the wind speed, is at a direction of 45° to the wind, but with increasing depth the magnitude and direction of the current both change markedly. With average winds, at a depth of somewhere around 100 meters (called the *depth of frictional influence*, *D*), the theory predicts that the current decreases to about 4 percent of its surface value and is directed at 180 degrees from the surface direction. Within recent years data has become available which indicate that the Ekman spiral does exist in the ocean (see Figure 9-21). However, if the water depth is less than the depth of frictional influence, the tendency is for the water in the entire water column to go in the direction of the wind so that in

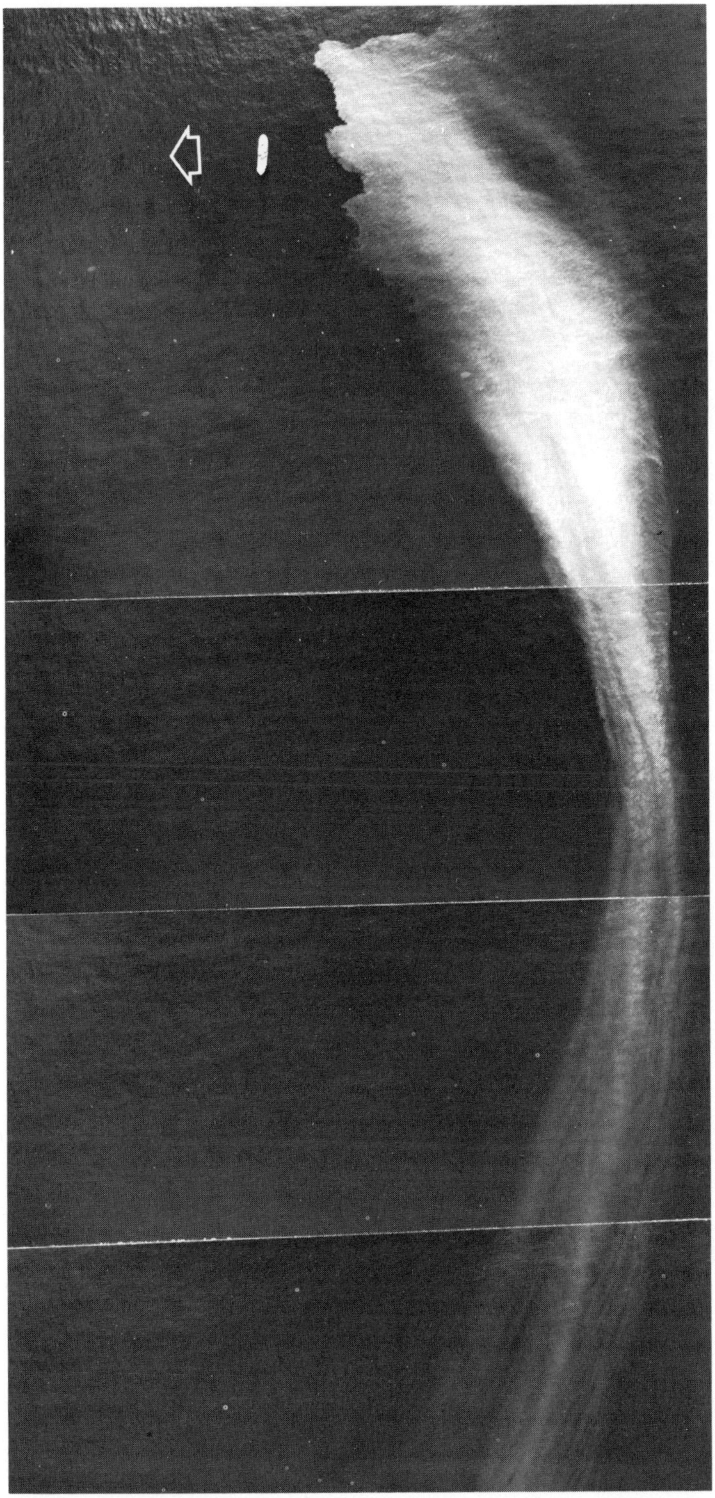

Courtesy: Richard Linfield, CBI, Johns Hopkins University

Figure 9-21 Aerial Photograph of Ekman Spiral Illustrating Dye Settling in Clear Water (Arrow Points North; Wind is Westerly)

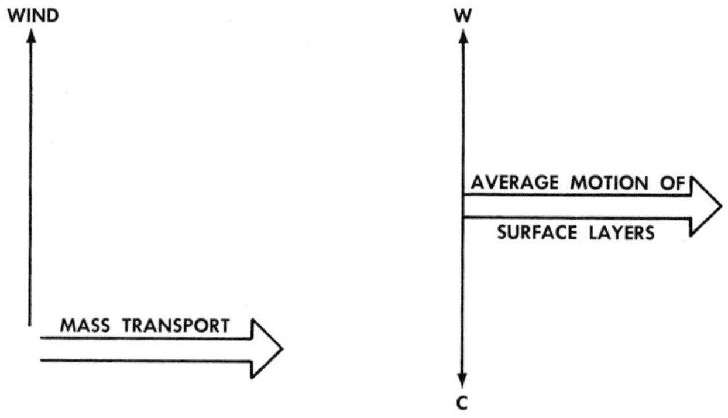

Figure 9-22 Mass Transport of Surface Layers Produced by the Ekman Spiral

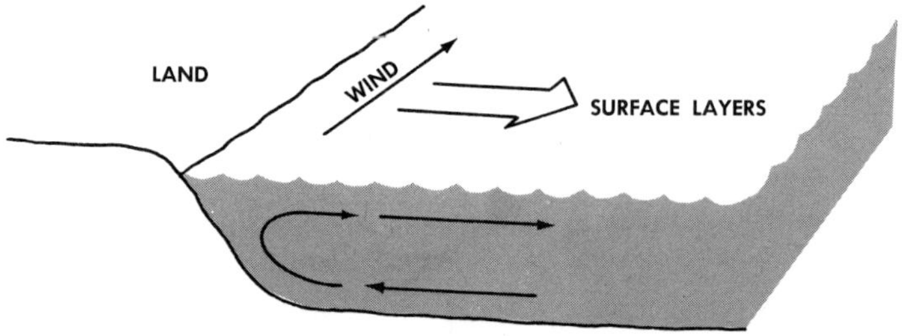

Figure 9-23 Upwelling Produced by a Longshore Wind

shallow coastal and estuarine waters, the Ekman spiral theory does not hold and the water is simply pushed in the wind direction.

If the motion of all the water in the layer down to the depth *D* is integrated, the average of all these motions (*mass transport*) is a motion at right angles to the wind. In other words, the average motion of the surface *layers* is at 90° to the wind. A vector diagram may be drawn to illustrate this result.

With the wind in the direction indicated and the resulting mass transport of the layer of frictional influence to its right, coriolis force must be to the right of this motion in the Northern Hemisphere. Note that the vector diagram shows a balance of forces between wind-stress force and coriolis force, which is the initial assumption made by Ekman. A practical application of this wind-induced motion of the upper layers is shown in some deep coastal areas when a wind will induce water to rise from subsurface layers by blowing along the coast rather than away from it. This *upwelling*, as it is called, is sketched in Figure 9-23. Where longshore prevailing winds exist in certain portions of the world, resulting upwelling continually supplies nutrients to the surface layers. This nutrient enrichment allows the development of fish populations as discussed in Chapter 14.

The Ekman spiral is not limited to the surface layers of the ocean but also is found near the ocean bottom and in the atmosphere near ground level. On the ocean surface, the wind exerts a drag force on the water with the surface motion ending up at 45° to the right of the wind-drag force.

In the case of both near-bottom water and near-ground air, the drag force is exerted by the stationary topography. Thus the drag force is exactly *opposite* to the fluid flow direction. This results in an inverted Ekman spiral for both near-bottom currents and near-ground winds.

916 Fluid Flow Summary. In the preceding paragraphs four idealized motions have been examined: *geostrophic flow*, *gradient flow*, *frictional flow*, and the *Ekman spiral*. The governing force balance relationships (stated vectorially) are summarized:

Geostrophic flow	$P = C$	(4)
Gradient flow	$P = C \pm S$	(5)
Frictional flow	$P = C + F \pm S$	(6)
Ekman spiral	$W = C$	(7)

917 Fluid Flow in Nature. The actual flow patterns which occur in nature may now be compared with the idealized representations already discussed. Starting at upper regions of the troposphere, the existing flow very closely corresponds with gradient or geostrophic flow. With few exceptions gradient flow is a very common occurrence. The flow is parallel to the isobars and actual, measured, wind speeds correspond very closely to wind speeds calculated from a knowledge of the pressure gradient and the latitude.

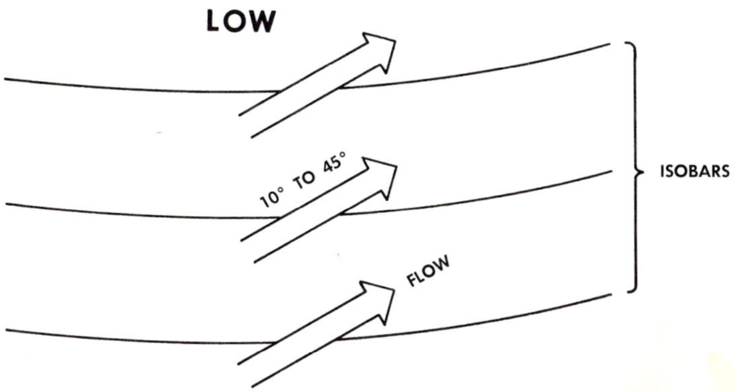

Figure 9-24 The Effect of Friction on Wind Direction

Close to the ground, frictional effects become pronounced. Within the lowest few thousand feet of the atmosphere, the friction of the ground has the effect of producing a change in direction of the wind vector toward low pressure. Winds blow across the isobars toward the low-pressure region making an angle with the isobars of about 10° over water and as much as 45° over land.

As indicated previously, the wind effect on the water surface produces the Ekman spiral which, in general, is a very close approximation to that which actually exists in nature. Instead of finding surface currents at an angle of 45° with the wind, the observed deviation is closer to 30°. This deviation may be produced by an additional surface current not associated with the Ekman spiral, but possibly associated with the production of surface waves (Chapter 19).

However, when an attempt is made to describe the surface and subsurface currents in the hydrosphere, which are a result of density distributions, the correlation between the actual flow and geostrophic or gradient flow does not hold as well as it does in the atmosphere. Nevertheless, for many practical applications, such as the prediction of iceberg drift in the North Atlantic, the geostrophic equations are utilized as a matter of course. Probably about 60 percent of the currents found in the ocean represent geostrophic or gradient flow; any large discrepancies found are usually the result of hydrospheric turbulence. Much work is being done on turbulence at this time and it is hoped that this will result in a better mathematical description of motion in both the atmosphere and the hydrosphere. This should make possible more accurate predictions of many variables in both fluids.

Additional Reading

Barry, R. G., and Chorley, R. J., *Atmosphere, Weather, and Climate*, Holt, Rinehart, & Winston, 1970.

Blair, T. A. and Fite, R. C., *Weather Elements*, 5th ed., Prentice-Hall, Inc., 1965.

Bolin, B., (Ed.) *The Atmosphere and Sea in Motion*, Rockefeller Institute Press, 1959.

Byers, H. R., *General Meteorology*, 3rd ed., McGraw-Hill, Inc., 1959.

Defant, A., *Physical Oceanography*, Vol. I, Pergamon Press, 1961.

Dietrich, G., *General Oceanography*, John Wiley & Sons, 1963.

Donn, W. L., *Meteorology*, 3rd ed., McGraw-Hill, Inc., 1965.

Duxbury, A. C., *The Earth and Its Ocean*, Addison-Wesley Publishing Co., Inc., 1971.

Eckart, C., *Hydrodynamics of Oceans and Atmospheres*, Pergamon Press, 1960.

Ekman, V. W., *On the Influence of the Earth's Rotation on Ocean Currents*, Almquist & Wiksells Boktryckeri Ab, 1905 (Reprinted 1963).

Haltiner, G. J., and Martin, F. L., *Dynamical and Physical Meteorology*, McGraw-Hill, Inc., 1957.

Hess, S. L., *Introduction to Theoretical Meteorology*, Holt, Rinehart, & Winston, 1959.

Hill, M. N., (Ed.) *The Sea*, Vol. I, Interscience Publishers, 1962.

McDonald, J. E., "The Coriolis Effect," *Scientific American*, May, 1952.

McLellan, H. J., *Elements of Physical Oceanography*, Pergamon Press, 1965.

Neumann, G., and Pierson, W. J., *Principles of Physical Oceanography*, Prentice-Hall, Inc., 1966.

Petterson, S., *Introduction to Meteorology*, McGraw-Hill, Inc., 1958.

Riehl, H., *Introduction to the Atmosphere*, McGraw-Hill, Inc., 1965.

Robinson, A. R., (Ed.) *Wind Driven Ocean Circulation*, Blaisdell Publishing Co., 1963.

Stewart, R. W., "The Atmosphere and the Ocean," *Scientific American*, September, 1969.

Stommel, H., *The Gulf Stream*, University of California Press, 1958.

Von Arx, W. S., *An Introduction to Physical Oceanography*, Addison-Wesley Publishing Co., Inc., 1962.

Weyl, P. K., *Oceanography*, John Wiley & Sons, 1970.

CHAPTER TEN

Wind Systems: Large and Small

1001 Introduction. In the winter of 1959 a dozen naval aviators sat dejectedly in the flight planning room of a naval air station in Texas. Outside, the weather was WOXOF (pronounced *walks-off*, a term derived from the weather encoding sequence for ceiling and visibility zero in fog and generally used in the aviation community to indicate the worst flying conditions). The reported tops of the cumulonimbus thunderheads were reported at 60,000 feet, and icing was severe—generally, a good day to stay on the ground.

One intrepid aviator came in, heard the state of the weather, and said, "Well, it's no sweat if I can get 'On top' (the term used for, above the clouds). I wonder what the winds are at flight level 850 (85,000 feet)?" This typifies the advent of the "real" jet age. Since that time the U-2 aircraft has demonstrated that the realm of aviation and space-flight will span the entire gamut of environmental phenomena. Significantly,

Waterspout in the Mediterranean
Courtesy: A. W. Berger, Great Neck, New York

it was the U-2 which brought back much information about an area of the environment where data was conspicuously sparse.

Only 15 years earlier had aircraft first encountered the high-velocity rivers of air above 35,000 feet. The Army Air Force B-29's in World War II were the first aircraft to fly high enough to verify the existence of the jet stream. The pace set by the aviation industry has been one that required diligence and increased research on the part of meteorologists. Information on the medium used is essential to any man who flies. A consideration of the atmospheric circulation is necessary for a total understanding.

1002 Tricellular Theory. As early as Aristotle's time, the dependence of the air movement upon the sun was postulated. Indeed, the realization that heated air ascends, and cooled air descends led to the development of the first theory on the general circulation of the atmosphere. Two in particular, the *tricellular theory* and the *eddy theory*, have enjoyed popularity.

It was pointed out in Chapter 5 how the rate of heating over the earth is far from uniform. There exists a large source of heat near the equator, while a sink for heat persists at the poles. The *tricellular theory*, based upon heat convection principles, was developed prior to the time that any extensive investigation of the upper atmosphere was done, and has served as the classical theory for many years.

Consider first a non-rotating earth, uniform of surface.* Over half of the area of the earth lies between 30°N and 30°S. This is the heat source area, where a surplus of solar radiation is received. Near the equator, the intense heat causes air to rise. The rising air creates a void. This void is compensated for by surface air being transferred from the poles. This works much the same as a heat engine. See both sketches in Figure 10-1. The room depicted has a radiator at one end. The air, warmed by the heat source, rises and is replaced by the cool air from the other end of the room along the floor, where its greater density dictates it be. In much the same way, the unicellular flow pattern on a non-rotating earth might be formed as shown.

The tricellular theory postulates that high-level motion would progress poleward to only 30° latitude. At this point the air would have sufficiently cooled to subside to lower levels. This is an area verified by measurement as a band of high pressure. Here the descending winds would split horizontally at the surface moving equatorward and poleward. That branch returning equatorward would complete the first cell of the three-cell motion.

* Uniformity of surface implies that no uneven heating exists due to differences of albedo and specific heat capacity.

The second cell would be formed by the poleward-moving surface air at 30°. This air would encounter the very cold air moving equatorward having originated at the poles. At about 60° they would converge horizontally, causing rising air. This rising air would then split at upper levels, going both poleward and equatorward horizontally, completing the other two cells. The picture is shown in Figure 10-2.

To further approach the existing conditions, the world must now be set in motion. When this happens, coriolis force comes into play. All of the previously described air motions are now deflected to the right in the Northern Hemisphere and to the left in the Southern Hemisphere

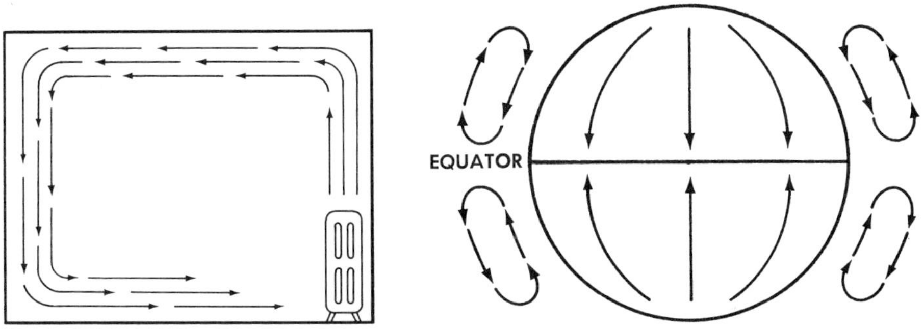

Figure 10-1 Convection

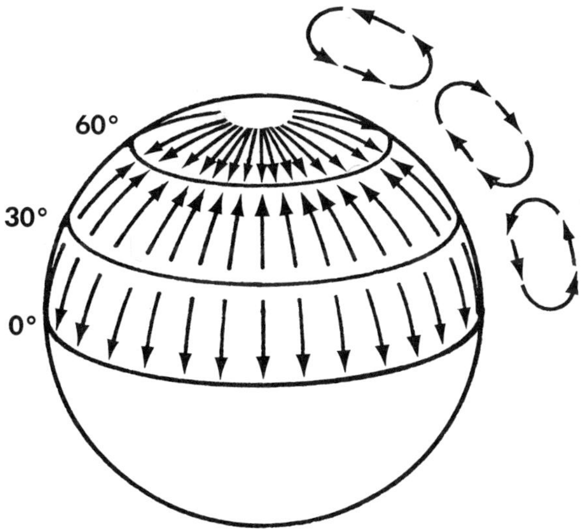

Figure 10-2 Simple Three-Cell Convection

(Figure 10-3). From this point, attention will be directed only to the Northern Hemisphere unless otherwise indicated since motion is similar in both areas.

The area between the equator and 30° is characterized by winds which flow from the northeast called the *northeast trades*,* so named because of their importance to the early commerce of the world.

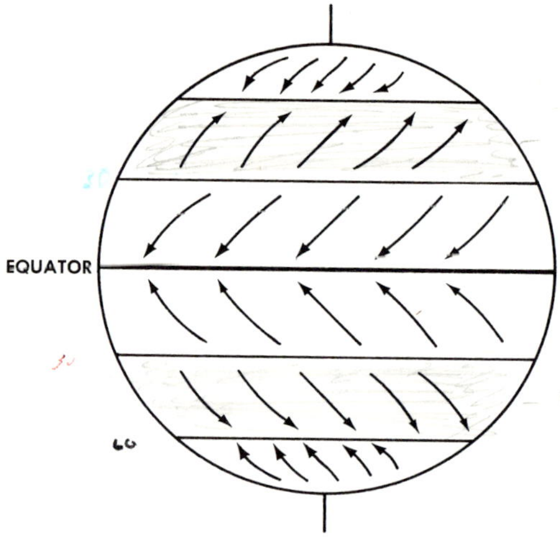

Figure 10-3 Tricellular Theory with Coriolis Effect

The equatorial area is known as the *doldrums* and also as the *intertropical convergence zone* (ITCZ). This is the area of convergence of the northeast and southeast trades, an area dreaded by the sailing men of old. The winds are extremely light, the seas glassy, and the humidity and heat oppressive. Many a sailor has been becalmed in the doldrums for long periods of time before auxiliary power was available.

The high-pressure area, centered near 30 degrees, is the familiar *horse latitudes*, noted for light winds due to the subsiding air. The name is said to have been given in the seventeenth century, because ships transporting horses to the New World, having been becalmed and running low on food and water, sometimes jettisoned some of their live cargo in this region. The area of the sub-polar low pressure at 60° will be given special attention later, since its influence is so widespread.

The tricellular theory has served the student of meteorology for many years and deserved mention. It should be pointed out that certain

* Winds are always named in terms of the wind source, i.e. a south wind blows from the south. Currents, on the other hand, are named after the direction of flow, i.e. a southerly current moves water toward the south.

observed facts are inconsistent with this theory, the most damaging being involved with upper-air motion. High-altitude investigation does not verify the upper flow as stated by the tricellular theory. The existence of the jet stream, a westerly flow aloft over the westerly flow at the surface in mid-latitudes, casts serious doubt on the entire theory.

1003 The Eddy Theory. The most recent concept of general circulation is based, not upon convection principles, but upon the necessity to conserve *angular momentum*. For a point mass, angular momentum may be represented by the product of the mass of the body involved, its speed, and the radius of curvature of its path (mvr).

Consider now that the earth, uniform of surface, is rotating, and its envelope of atmosphere is rotating with it, motionless with respect to its surface. In this instance each parcel of air has its own angular momentum which must be conserved, i.e. the value must remain unchanged. In order to conserve angular momentum with an unchanging mass of the air parcel involved, any change in speed must be accompanied by a compensatory change in the path radius of curvature, or vice versa. Observe the area between the equator and 30°N where the winds are from the northeast. Compared to a particle on the surface of the earth near the equator, which has a linear speed of 900 knots in an easterly direction, the air in the trades is traveling at a rate less than this, since it is moving toward the west with respect to the earth's surface. Therefore a deficit of angular momentum must exist between the equator and 30°N, so that angular momentum must have been transferred to another area of the atmosphere which has motion.

Investigation of the area of mid-latitude westerlies shows that air in this region is apparently blessed with a surplus of momentum since it is traveling at a higher speed than the surface area it overlies. A similar analysis can be done in the area of the polar easterlies.

There are three areas (*trades, westerlies, polar easterlies*) which have an imbalance in momentum (see Figure 10-3). Something must occur to allow a transfer between them to restore global equilibrium. Here is where the eddy comes in. The eddies are the familiar high- and low-pressure systems which migrate from one area to the other, rotating clockwise and counterclockwise respectively. In the process, they transfer angular momentum assisting in the balance between the wind regions. At the upper level there exist wavelike motions, instead of closed clockwise and counterclockwise circulations, which aid in this transfer of momentum. In fact, preliminary computations attribute the major momentum transport to these high-level waves.

This theory is more in consonance with observations taken at high levels and with theoretical mathematical calculations made on elec-

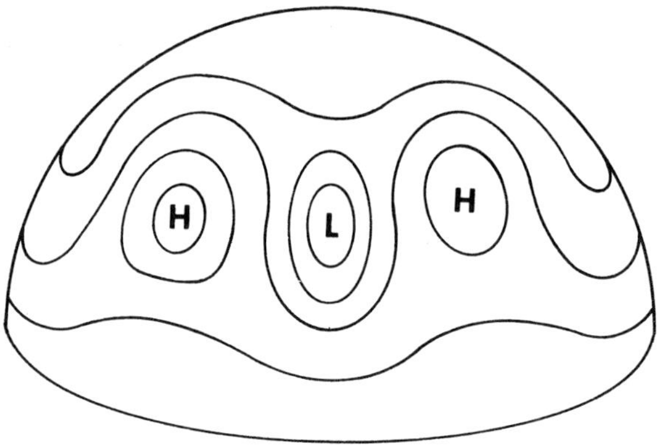

Figure 10-4 Migratory Closed Circulations

tronic computers. In fact, strikingly similar reproductions of this effect have been demonstrated in the laboratory.

To put the picture that has been constructed into final perspective, consideration must now be given to the effect of the heterogeneous surface of the earth. The changes in the circulation pattern are a function of the various surfaces which choose to reject or accept, in a qualified fashion, the sun's incoming energy. This results in an alteration of the strict latitudinal placement of the high- and low-pressure bands. In addition, the heating differences give rise to gradations of the pressure within the belts characterized as high and low. In other words, there are areas in the low-pressure belt that are significantly lower in pressure and there are areas in the high-pressure belt that are significantly higher than other parts. A discussion of these anomalies in the pressure system follows. These standard pressure systems are based on the seasonal mean, although daily and weekly variations do occur.

This discussion will be broken down into a seasonal treatment. Winter in the Northern Hemisphere will be discussed first. The Northern horse latitude, an area of high pressure, is identified by two high-pressure, closed circulations, one over the Atlantic and one over the Pacific. The Pacific high (anticyclone) is the more persistent. The pressure pattern over the Atlantic has a tendency to merge with the continental high pressure over North America and North Africa during the winter. A third very intense high-pressure cell exists over Asia. It is centered about 45° north and shows the tendency of the pressure belt to migrate in response to the intense cold that exists over Asia in the winter. The subpolar belt of low pressure essentially divides itself into two well-defined circulation cells. They are the Aleutian low and the Icelandic low. The ITCZ (doldrums) shows a tendency to migrate farther south, toward the Southern Hemisphere summer. The Southern Hemisphere

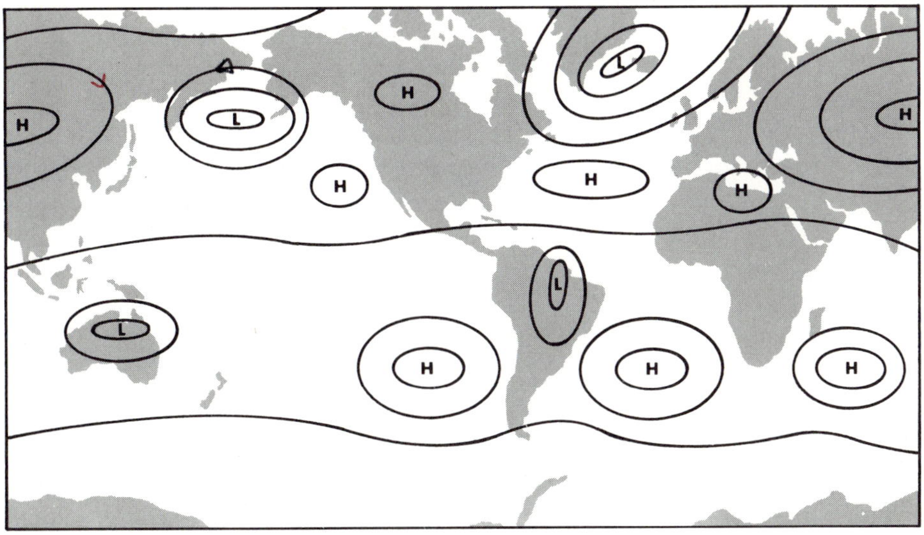

Figure 10-5 Mean Pressure Pattern (January)

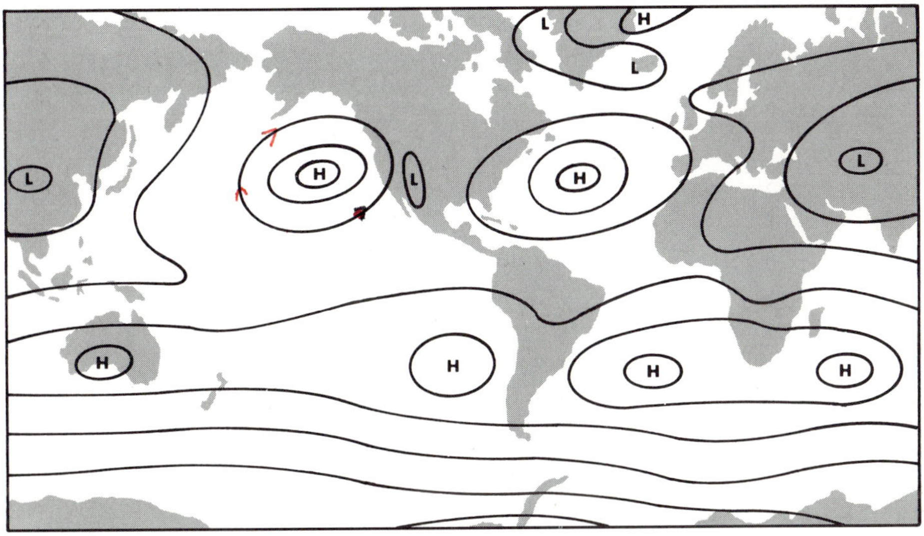

Figure 10-6 Mean Pressure Pattern (July)

picture during this period (January) has a more regular pattern. Three high-pressure cells in the ocean areas about 35° south and low-pressure cells over South America, Australia, and South Africa, can be identified.

Now consider the Northern Hemisphere's summer (July). In general, the existing circulations over the water areas near the equator are intensified, because their temperatures remain cool compared to the surrounding land areas. However, the closed high-pressure circulations have turned into thermal lows over North Africa, Asia, and the southwest United States. In the region of the subpolar low, the Icelandic low is discernible but weak, and the Aleutian low cannot be identified as a closed system. Again, during this half of the year, the belt of convergence near the equator (ITCZ) is seen to migrate toward the summer hemisphere. The migration of this equatorial trough, as it is sometimes called, follows the inclination of the sun, and is modified by the placement of the continents amidst the vast ocean area.

1004 Upper-Air Motion. The discussion so far has described the surface circulation. How does the movement of air vary as altitude increases? This situation can be categorized as follows. The area between about 15°N and 15°S is easterly in nature from the surfacc to the tropopause, achieving an average maximum value of about 20 knots. Aloft, over the remaining regions (excepting the poles) the flow is westerly. The speed of the winds gradually increases with height in mid-latitudes reaching a maximum value near the tropopause. It is in the region over the middle latitudes of the Northern Hemisphere that the spectacular phenomenon known as the *jet stream* exists. The existence of this river of rapidly moving air was suspected prior to 1940. The maximum effort by meteorologists in World War II promoted extensive upper-air soundings which resulted in identifying the jet. Here, too, the discovery was verified by long-range aircraft reporting winds in excess of 200 knots at altitudes near 35,000 feet.

The *jet stream* is defined by the World Meteorological Organization as a ". . . strong narrow current, concentrated along a quasi-horizontal axis, in the upper troposphere, or in the stratosphere, characterized by strong vertical and lateral wind shears, and featuring one or more velocity maxima." Normally the dimensions of this stream are thought of as thousands of kilometers in length, hundreds of kilometers in width and "some" kilometers in depth. For terms of definition the lowest velocity considered to be of jet-stream speed is 50 knots. The velocities of the jet stream show a seasonal variation, being much stronger in winter than summer. Speeds of 200 knots over the United States in the winter would be common, while in the summer they might be more in the 75–100 knot range. Verified speeds of 250 knots have been measured. Although a 300-knot maximum has been frequently quoted, it has not been verified.

It is quite often the case that the jet splits and can be shown as two separate bands of high-speed air about 500–800 km apart. The predominant one is called the *polar front jet stream* and is associated with the polar front. The undulating character of the polar front is characteristic of the circulations aloft, in general.

The pressure pattern aloft demonstrates an aversion to closed circulation systems and seems to consist of one closed global circulation. The existence of troughs and ridges in the isobaric pattern typifies the upper atmospheric pattern.

Various reasons have been proposed explaining the maintenance of the jet stream; prominent among them is the temperature structure. There appears to be a close connection between the upper wave pattern (i.e., the polar front) and the movement of the jet stream.

The second jet identified is the *subtropical jet stream*, normally located between 20° and 30°N. These two jet streams are part of the main jet stream which averages between 35,000 and 40,000 feet in height.

The upper atmosphere contains another stream of rapidly moving air centered near the North Pole. The appearance of this moving air is predicated on the long polar night. This wind system is called the *polar night jet*. Its core is centered in the vicinity of 80,000 feet. It is of a lesser speed than the polar jet and its duration is only from November to March. During the balance of the year, light easterly winds prevail aloft, accompanying the polar easterlies at the surface.

The navigational characteristics of these jets have created many problems for operations. Since air traffic is becoming so dense, the assignment of altitude and airway must be specified. In every instance possible, allowances are made for high-altitude aircraft, to enable them to utilize these winds. The information available to the pilot can aid him in requesting that his aircraft be assigned to the most advantageous altitude, appropriate to his direction of flight. It is little wonder that some of the foremost work on jet-stream analysis is done by the meteorologists employed by the major airlines. One airline on the Tokyo to Honolulu route saved a half-million dollars over a series of 220 flights by utilizing jet-stream techniques. It is obvious that pilots of high-performance aircraft should have a knowledge of jet-stream identification.

1005 Clear-Air Turbulence (CAT). Another phenomenon associated with the jet stream is *Clear-Air Turbulence*.* There appears to be a definite relationship between the *shear*† of the velocity present in jet streams and the frequency of occurrence of CAT. What makes CAT particularly hazardous for the pilot is the fact that he is probably unaware when he will encounter it next and the *flight configuration*‡ of the aircraft may not be at the optimum to sustain it. Turbulence can be encountered at any level, but the rough air associated with convection

* *Turbulence* may be defined as motion exhibiting rapid, irregular fluctuations.

† *Shear* is defined as the change in velocity (speed and direction) with distance measured in a direction perpendicular to the flow.

‡ A general descriptive term encompassing speed, control surface placement, and power settings.

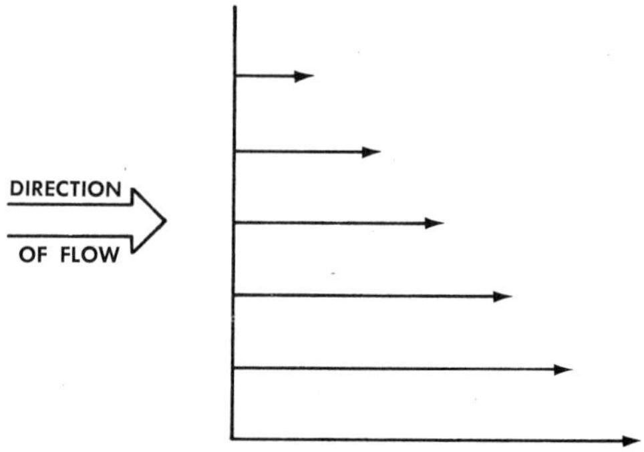

Figure 10-7 Vector Shear of the Wind

currents is normally identified by the visual signs, towering cumulus and cumulonimbus. No pilot would penetrate a squall line unless his aircraft were in the proper flight configuration to deal with it. Structural damage can be avoided or minimized and maximum flight stability can be sustained if proper techniques are used.

Although information on CAT is incomplete, some generalizations can be made that enhance safe flight. There appear to be two general zones of CAT near the *jet core*. The first is below and to the left (back to the wind) between the altitudes of 22,000 and 28,000 feet. This is an area of *maximum wind shear*. The second area is localized just above the jet core between 35,000 and 50,000 feet. This area extends above the core and to the right for a considerable distance (on the order of 200 miles). Naturally, in the core itself, there is little shear (by definition), and consequently no CAT. CAT appears to exist in patches, horizontally, varying in both time and space.

1006 Monsoons. One circulation feature of the atmosphere is peculiar to a single geographical area of the globe. This is the *monsoon circulation*. The term *monsoon* means seasonal wind (Arabic, *mausin*—a season). This circulation is tied directly to the ability of the land areas to cool quickly in the winter, and to warm just as quickly in the summer. The enormous mainland of Asia is subject to these fluctuations. When winter comes to Siberia, temperatures become intensely cold. The unofficial record for low temperature in the Northern Hemisphere is held by Oimekon, Siberia, at −108°F. As one would expect, this causes severe subsidence of the air that overlies the land. This high-pressure area produces outflow of air at the surface, coriolis turning it to the right, and a flow of air from land to sea.

In the summertime, this process is reversed causing a low-pressure cell to form. This moves the surface air inward from the sea, toward the land. This air, which has picked up great amounts of moisture over the water, is now lifted over the land surfaces of India, on the way up the slopes of Nepal, Tibet, and China. The reaction that takes place is notorious. The rains that fall from this rapid and widespread condensation begin in April and May in Burma, followed by the onset in India in late May and early June. The economy of the area, living habits of the people, and the motion of the sea are attuned to the arrival of these winds and rains. As a consequence of this, Cherrapunji, India, has the reputation of being the wettest spot on earth, not covered by an identifiable body of water: its record rainfall in one year is 1,042 inches, its record rainfall in one month is 366 inches, and its record rainfall for a five-day period is 150 inches (12.5 feet).

As a result of the monsoon, the currents of the ocean respond with a seasonal reversal. The Indian monsoon current exhibits this behavior. This reversal of wind and current with the change of season accounted for the Arabian trade with Africa and India. Their dhows sailed the Arabian Sea from October to May on winds from the northeast and sailed from June to September on the southwest winds. This is the only large-scale instance of this occurrence in oceanic circulation. The Davidson current off the coast of California shows a periodic appearance which gives the impression that the current reverses, but this is not the case. It is the emergence from depth that causes the Davidson current to flow northward seasonally.

Monsoonal characteristics have been attributed in a lesser degree to other areas of the world such as Spain, Africa, the United States' Gulf coasts, and Chile. It is normally difficult, however, to separate the monsoonal wind component from the other winds of a random nature which characterize these areas.

1007 Local Winds. Although not of large-scale significance, certain local winds are of some importance. The *land-sea breeze* is a typical heat engine phenomenon. This is the result of the juxtaposition of two bodies of differing thermal characteristics. The principles involved are similar to the monsoon situation with the exception that a shorter time frame and smaller distances are involved. The rising of the sun begins the period of *insolation*.* The land area warms much more quickly than the adjacent water area and the subsequent rising of the warm air begins. The rising air is replenished at the surface by that coming in from sea. The air which returns aloft to sea finally cools, sinks, and enters the cycle again.

With the coming of nightfall, the land then cools to a lower temperature than the adjacent water. The air subsides over the land and is

* *Insolation* is the solar energy received per unit area per unit time at the earth's surface. It is usually expressed in terms of calories/cm^2/min (Langleys/min).

forced out over the water where the cycle reverses the daytime situation. The onset of the sea breeze (remember, it is named for where it originates) is usually felt about mid-morning. As soon as the temperature contrast swings the other way after nightfall, the land breeze begins. This effect is one of the reasons that many seacoasts generally have mild temperatures. The cool, moist winds are a welcome relief to the seashore dweller on a summer afternoon.

This phenomenon is also an important clue to the sighting of land while at sea. The convection currents set up over land, even small islands, will cause the formation of tell-tale cumulus clouds. Since the clouds are formed at altitude, they are visible above the horizon and pinpoint the nearest land. This has saved many a neophyte navigator from embarrassment. Furthermore, such basic knowledge has saved many lives in survival situations. Information of this nature should be part of the basic knowledge of any navy man, shipborne or airborne.

Throughout the world, special winds exist as the result of local circumstances, mainly of topography. Others are the result of the general

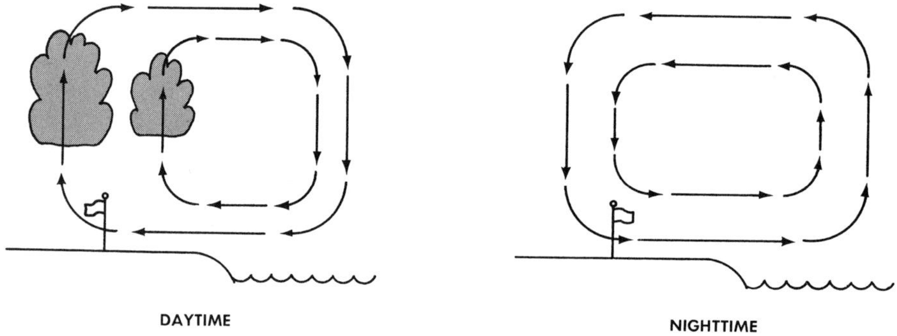

Figure 10-8 Seabreeze-Landbreeze

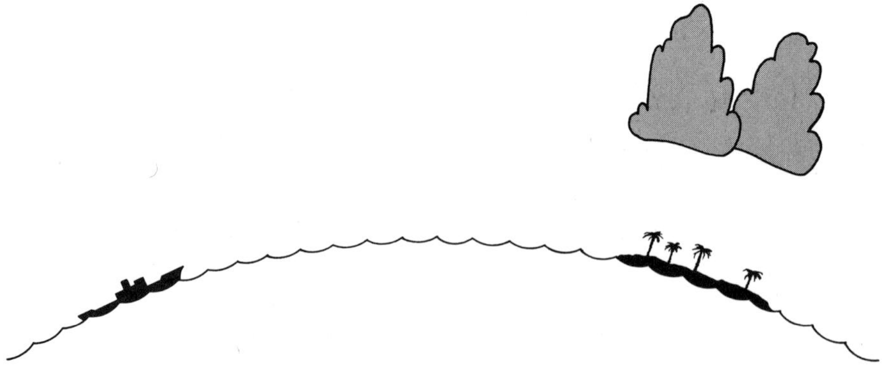

Figure 10-9 Island-Formed Cumulus Clouds

circulation. These are normally given names by the local populace, gaining permanence through repeated usage. Quite often they fall into a few general categories, two of which are: *katabatic* (mountain) winds, and *anabatic* (valley) winds. Anabatic (Greek, *anabasis*—moving up) winds are usually formed when one wall of a valley is heated by the sun to a degree greater than the shaded valley itself. Air will rise where the heating takes place. A new supply of air will be provided from the valley floor and moves up the slope of the valley sides. After the heating process has stopped, the reverse effect will begin. This is the katabatic (Greek, *katabatos*—moving down) or mountain wind.

Normal gentle breezes are taken for granted, but where local topography accents the process and provides stronger than usual winds, they are given a particular name. *Drainage winds* make up another category of special winds. These are usually the result of an accumulation of cold stagnant air on a plateau, perhaps encouraged by a basin type shape of the surface. This large accumulation of cold air, is spilled down to lower levels by the force of an advancing pressure center. At times an adjacent valley will help to channel these winds creating high speeds. The Mediterranean is noted for some of these winds. Any Southern Californian is familiar with the Santa Ana winds that blow to the coast out of the Santa Ana and other nearby canyons. These winds are strong, extremely dry, and often carry an annoying burden of fine sand from the desert region. This wind can raise, adiabatically, the normally pleasant temperatures (about 63°F day-night mean) of San Diego to over 100°F—a radical change for that part of the country.

Another wind that falls under the category of general circulation, but with a special twist, is the *Foehn*. This is the result of winds passing over an orographic (Greek, *oros*—mountain) barrier, such as the Rocky Mountains, cooling and condensing moisture on the way up the slope and then warming on the way down the slope. In its upward travel, the cooling is at the moist adiabatic rate (after the condensation level is reached), and, on the way down, the warming is according to the dry adiabatic rate. Consequently, final air temperatures are markedly higher than initial ones. These winds often are found on the lee side of the Rockies in Wyoming and Montana. Here they are called *chinook* winds. At Havre, Montana, a rise in temperature of 31°F was recorded in three minutes, at the onset of a chinook. This wind had warmed 5.5°F/1,000 feet in its descent of about 5,000 feet. Note that 5.5°F/1,000 feet is the dry adiabatic lapse rate.

1008 Tornadoes. Tornadoes are violently rotating (usually cyclonic) columns of air having extremely high wind speeds. These may easily reach velocities of 300 knots in the center. They are created by extremely unstable conditions, and are impossible to predict specifically. The general conditions which must be present, however, can be observed in time to issue tornado watches or warnings.

Tornadoes have been statistically connected with cold-front convective activity. The most active region in the world is the Mississippi Valley because of the spring and summer heating over the great western plateau.

The tornado funnel has the diametrical expanse of a mile or less; it is recognized as a dark pendant funnel-shaped cloud reaching down to earth from the heavy thick clouds above. It is usually part of a larger low-pressure "tornado cyclone"* which is about ten miles across with average winds of fifty knots. The funnels are easily tracked by radar once they have been positively identified and reported.

Waterspouts are tornadoes over water which draw moisture upward just as dust is raised over land. A true spout is cloud-connected just as is a tornado. Others have been seen to rise up from the water in much the same manner as land-borne "*dust devils.*" Waterspouts are comparatively rare as contrasted to tornadoes because the unique thermal properties of water exert a moderating effect on instability and prohibit the existence of large air-sea temperature differences.

1009 Instruments. Various devices are used to measure the direction and speed of the wind. They range from the most elementary to the very complex. The most familiar is the anemometer (Greek, *anemos*—wind). It consists of a de'ice which will align itself in the direction of the wind, as does the traditional weather vane, and a mechanism to change the motion of the air to that of another type. Usually, this apparatus is either a small propeller or a combination of three radially mounted cups which catch the wind and turn a very small electric-current generator. This arrangement facilitates remote readout of the information or conversion of the data to a time record. Figure 10-10 shows the most popular installation.

Another anemometer is the hand-held type shown in Figure 10-11. This is also the electric generator type and is used where a permanent installation is impractical.

One of the most elementary wind-indicating apparatus, the wind sock seen at all airports, shows wind direction and, to a limited degree, speed. The sock is so designed that a known minimum speed will cause it to extend to a predetermined horizontal angle. The most common type encountered will be fully horizontal at fifteen knots. While quite simple

* This is not to be confused with the extratropical cyclone discussed in Chapter 16.

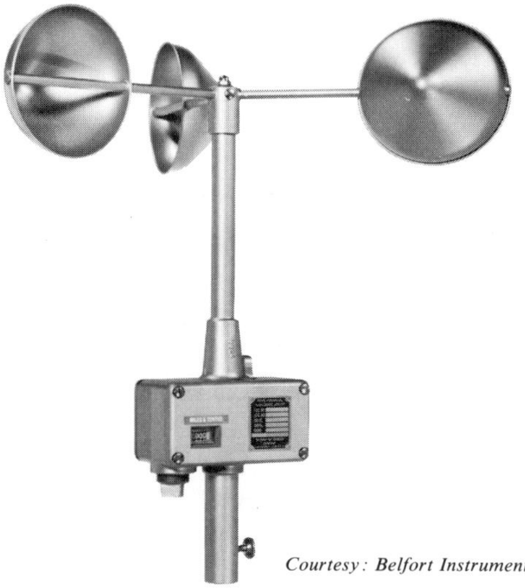

Courtesy: Belfort Instrument Co.

Figure 10-10 Three-Cup Anemometer

Courtesy: Belfort Instrument Co.

Figure 10-11 Hand-Held Anemometer

in concept, it remains a valuable aid to every pilot in that it is reliable, easily visible, and instantaneously available when most critically needed, on landings and takeoffs.

The principal means of measuring the direction and velocity of winds aloft is by tracking the movement of a balloon as it ascends. Since balloons are released on a routine basis, the balloon-borne radiosonde described in Chapter 4 is tracked upon launching and the data collected. At other times a balloon alone is released and tracked visually, by radio signals, or by radar. These are called PIBALs for pilot balloons or RAWIN for radar wind observation. Azimuth and elevation angles permit computations of direction and speed during the flight of the balloon at the known ascent rate.

1010 Safety. A ship at anchor is always vulnerable to strong winds. The stresses of the air-ocean interface can be extremely great. A conscientious mariner will be aware of the prevalent phenomena in the ports he will visit and their most probable time of occurrence. A little forewarning and preparation can go a long way in enhancing the safety of a ship in a strange port.

Additional Reading

Barry, R. G. and Chorley, R. J., *Atmosphere, Weather, and Climate*, Holt, Rinehart, & Winston, 1970.

Battan, L. J., *The Nature of Violent Storms*, Anchor Books, Doubleday & Co., Inc., 1961.

Byers, Horace R., *General Meteorology*, 3rd ed., McGraw-Hill, Inc., 1959.

Donn, William L., *Meteorology*, 3rd ed., McGraw-Hill, Inc., 1965.

Edinger, J. G., *Watching for the Wind*, Anchor Books, Doubleday & Co., Inc., 1967.

Haltiner, George J. and Martin, Frank L., *Dynamical and Physical Meteorology*, McGraw-Hill, Inc., 1957.

Huschke, Ralph E., *Glossary of Meteorology*, American Meteorological Society, Boston, Mass., 1959.

Matthews, Samuel W., "Science Explores the Monsoon Sea: Indian Ocean," *National Geographic*, October 1967 (with chart).

Petterson, Svere, *Introduction to Meteorology*, McGraw-Hill, Inc., 1958.

Reiter, E. R., *Jet Streams*, Anchor Books, Doubleday & Co., Inc., 1967.

Riehl, Herbert, *Introduction to the Atmosphere*, McGraw-Hill, 1965.

Starr, Victor P., et al., *Observational Studies of the Atmospheric General Circulation*, Massachusetts Institute of Technology, Cambridge, Mass., July, 1966.

——, "The General Circulation of the Atmosphere," *Scientific American*, December, 1956.

Sverdrup, H. V., Johnson, Martin W., and Fleming, Richard H., *The Oceans*, Prentice-Hall, Inc., 1942.

Tepper, M., "Tornadoes," *Scientific American*, May, 1958.

———, *Radio Weather Aids* (H. O. Pub. 118) Naval Oceanographic Office, Govt. Printing Office, Washington, D.C., 1963.

———, *Synoptic Patterns for Clear Air Turbulence*, NWRF 15-0965-107, Naval Weather Research Facility, September, 1965.

———, *The Upper Atmosphere*, NWRF 26-0665-106, Naval Weather Research Facility, June, 1965.

———, *Tornado*, Environmental Science Services Administration ESSA/PI 660028, U.S. Govt. Printing Office, January, 1967.

CHAPTER ELEVEN

Oceanic Surface Currents

1101 Introduction. Anyone having even an elementary acquaintance with the ocean knows of the great currents which exist at its surface. Some of these, such as the Gulf Stream in the North Atlantic and its counterpart in the North Pacific, the Kuroshio, have a profound effect on continental climate. This is usually considered to be beneficial since Great Britain and southwestern Alaska are both warmer than would be expected from latitudinal considerations alone. There are surface currents, however, which may produce harmful effects on the total environment.

One example of this is the Humboldt (Peru) Current which flows northward along the western coast of South America. Under normal circumstances there is a large amount of upwelling associated with the edge of this current providing nutrients for enough phytoplankton to support the largest anchovy population in the world. These fish make up one of the largest single industries in the Peruvian economy and

Courtesy: Martin H. Miller, Silver Spring, Maryland

account for the fact that Peru has led the world in fish catch for the last few years. Occasionally, however, the Humboldt Current changes its position enough to cause a cessation of the all-important upwelling. Not only do the anchovies disappear, causing the loss of untold millions of dollars to the Peruvian economy, but large numbers of other marine organisms die and decay resulting in hydrogen sulfide being released to the atmosphere.

This change in position of the Humboldt Current is called *El Niño* (after the Christ child because of its usual occurrence during the Christmas season), or sometimes *Callao Painter*, after the discoloring effects which are caused by the gases of decaying organisms.

Of course not all surface current systems are capable of producing a cataclysm of the order of magnitude of El Niño, but this does not make the knowledge of all surface currents any less important. Benjamin Franklin was aware of this when he produced the first chart of the Gulf Stream to aid American ships make faster crossings of the North Atlantic. But it remained for Matthew Fontaine Maury to initiate the scientific study of ocean currents about three-quarters of a century later.

Maury not only gathered together enough data to generate dependable charts, but he also tried to correlate these data in an attempt to ascertain the causes and variability of ocean currents. This work has continued to the present and is not complete, even today. However, enough information has become available to piece together a reasonably logical description of oceanic surface currents.

1102 Basic Causes. As in the case of atmospheric motion, one of the major causes of motion in the sea is uneven heating. However, the atmospheric flow pattern discussed previously is somewhat different

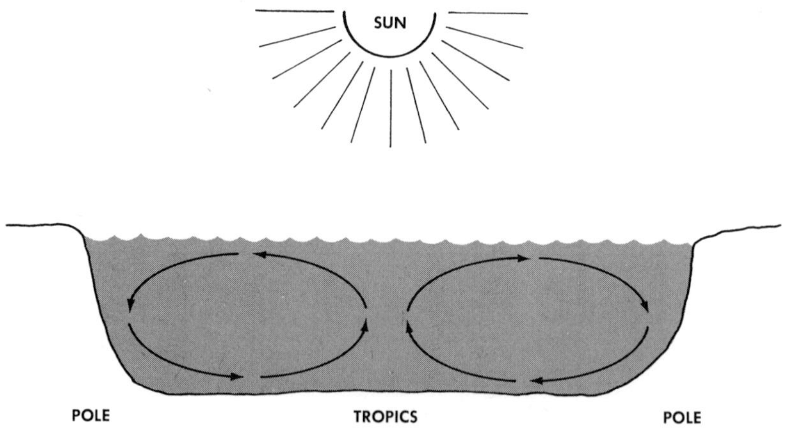

Figure 11-1 Idealized Thermohaline Flow in the Ocean

than that in the ocean, because in addition to the direct effects of uneven heating, there are two other important factors which must be taken into account. These are 1) wind (itself produced by uneven heating) acting on the water surface and 2) the containment of the oceans within the boundaries set by land masses. Due to the interference of land masses, no currents run all the way around the world except in the Antarctic region.

In actuality there are two basic systems which must be superimposed, one upon the other. The first of these is the system produced directly by uneven heating wherein the waters at lower latitudes are heated, become less dense, and spread out over the surface toward the poles. As they drift toward the poles these waters are cooled and finally sink. In this manner a giant convection cell is set up similar to the single cell atmospheric model, wherein surface water sinking at the poles flows toward the equator, rises in the equatorial regions, and flows away from the equator along the surface.

In addition to this basic flow poleward, the surface winds, combined with land mass placement, produce a different system. The resulting surface currents are a combination of these two flows. Since by far the greatest effect is due to winds, an attempt will be made to develop a model of ocean currents produced by wind forces and land placement alone. This will then be compared with what actually exists in nature.

1103 An Oceanic Current Model. As a start, the model assumes that the winds in existence are those described in the three-celled theory discussed in the previous chapter. This is a reasonable assumption since the tricellular model describes the *surface* winds quite well. As may be recalled, winds in the Northern Hemisphere are northeast in the lati-

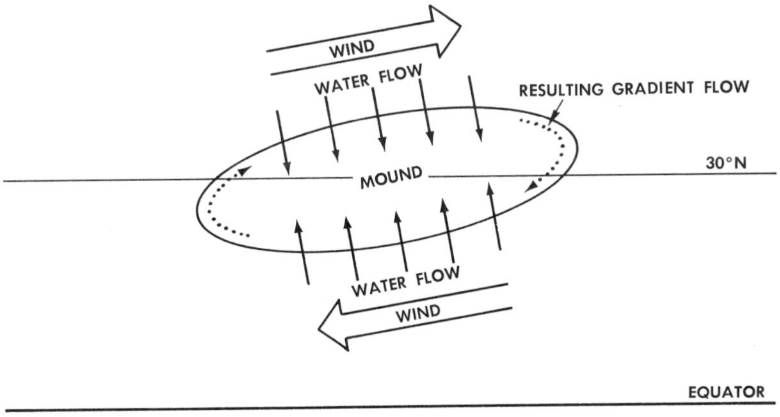

Figure 11-2 Production of an Oceanic High-Pressure Area at 30° by Prevailing Winds

tude belt from 0° to 30°, southwest in the belt from 30° to 60°, and northeast again in the belt from 60° to the pole, with a mirror image of this system in the Southern Hemisphere.

As may be recalled from the previous discussions of the Ekman spiral, when a wind blows there is a transport of the upper layers (about 100 meters thick) at 90° to the right of the wind in the Northern Hemisphere. The result is that, with a wind blowing from the northeast, the oceanic surface layers will be caused to move toward the northwest, which is the case between the latitudes of 0° to 30° in the Northern Hemisphere.

Similarly, with a southwest wind the upper hundred meters or so of surface waters are transported to the right and a southeast* flow develops. The effect of these two currents is to pile up water within a region centered somewhere around 30° latitude, as seen in Figure 11-2.

This mound of water piled up by these two wind-driven transports creates a high-pressure ridge at about 30° latitude. The water, under the influence of this pressure distribution and coriolis force, will produce geostrophic flow toward the southwest between 0° and 30° and toward the northeast between 30° and 60°. Since there are land masses on each side of the ocean, the water must go somewhere. As it completes its path, it tends to produce a current gyre in a clockwise direction about this high-pressure cell. Just as in the atmosphere, a clockwise rotation is found about a high.

A little farther north there is a northeast wind between the latitude of 60° and 90° which would cause the surface layer to move toward the northwest. Consequently at 60 degrees latitude water is directed toward

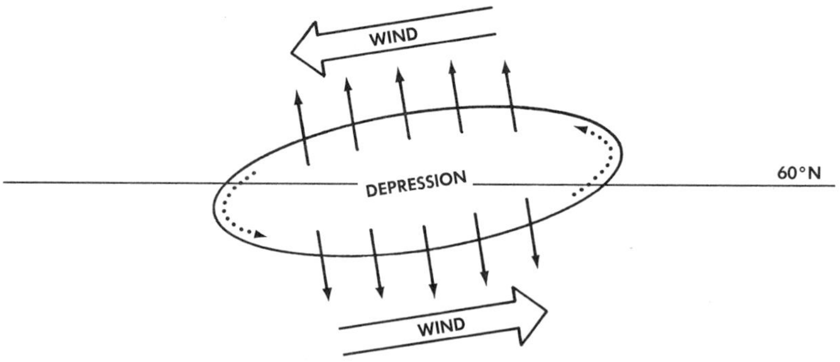

Figure 11-3 Production of an Oceanic Low-Pressure Area at 60° by Prevailing Winds

* Keep in mind that winds are named by where they have been, while currents are described in terms of where they are going.

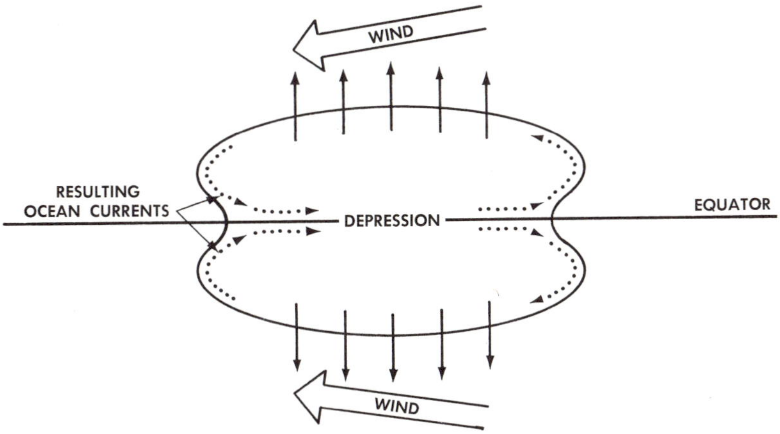

Figure 11-4 Production of an Oceanic Low-Pressure Area at the Equator by Prevailing Winds. (Note the two gyres produced from this single depression.)

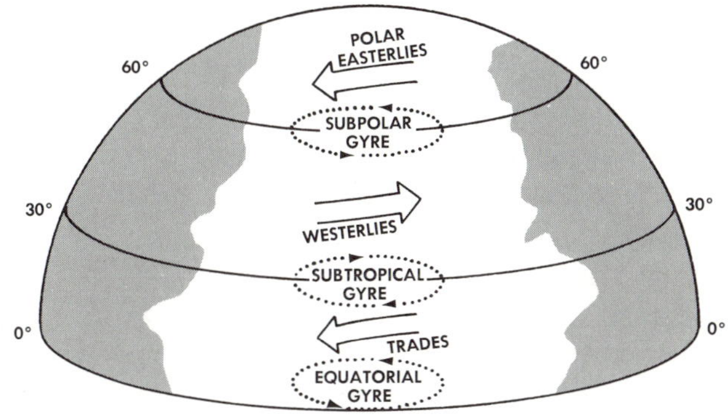

Figure 11-5 The Complete Model for the Northern Hemisphere Wind-Driven Currents

the southeast by lower latitude winds and at the same time toward the northwest by winds at higher latitudes. This results in a low-pressure trough at a latitude of about 60 degrees. Considering both the placement of the continents on each side of the ocean and what has been learned about low-pressure regions, the circulation around this low-pressure area will result in a current having a counterclockwise direction.

In equatorial regions, the situation may be treated similarly. The winds north and south of the doldrums produce surface-layer motion such that in both cases the water motion is away from the equator. In other words, close to the equator a low-pressure system is developed in the hydrosphere. Once again, counterclockwise rotation would be expected around a low-pressure system in the Northern Hemisphere, while clockwise flow would be expected in the Southern Hemisphere. Thus two gyres result since flow direction about a low is opposite in the different hemispheres.

The model is now complete. Using the tricellular wind theory, it has been shown that the resulting currents in the Northern Hemisphere should consist of a counterclockwise gyre close to the equator (*equatorial gyre*), a clockwise gyre north of that (*sub-tropical gyre*), and a counterclockwise one north of that (*sub-polar gyre*).

Similarly, about the same condition exists in the Southern Hemisphere, except that the gyres rotate oppositely, producing a mirror image. The *equatorial gyre* is clockwise in its rotation, the *sub-tropical gyre* counterclockwise, and the *sub-polar gyre* clockwise, due to the opposite direction of the coriolis force in the Southern Hemisphere.

1104 The Model vs The True Picture. It is now appropriate to compare a simple model with the actual currents existing in the world's oceans. In all of the world's oceans there is a *sub-tropical gyre;* both north and south of the equator this portion of the model appears to hold fairly well. In the Pacific Ocean the *north equatorial gyre* is well established, composed of an equatorial current and an oppositely moving equatorial counter current north of the equator. There is also a *south equatorial gyre*, displaced somewhat north of the geographical equator. This is not surprising from the position of the *intertropical convergence zone* and the *oceanographic thermal equator*, both of which are displaced north of the geographical equator.

In the northern oceans the subtropical gyre composed of the *Kuroshio system* in the Pacific and the *Gulf Stream system* in the Atlantic is also very well developed. In addition, the *Irminger current*, an offshoot of the Gulf Stream system,* combines with the *East Greenland current* to produce the sub-polar gyre. However, the sub-polar gyre in the Pacific is not so well developed, although the Alaska current tends to produce a flow of this type.

In both the South Pacific and South Atlantic there is some evidence of the existence of the sub-polar gyre. In the South Pacific, a polar current running from east to west close to the Antarctic Continent and the West Wind Drift (Antarctic Circumpolar Current) somewhat north of this in the opposite direction, make up the larger portion of the sub-polar gyre. The southern South Atlantic also exhibits very similar properties so that it appears the model fits the southern oceans quite well.

From an unrefined point of view this crude model fits the real ocean much better than might be expected from the simplicity of the initial assumptions. It appears that some of the differences between the model and actual current patterns may be explained on the basis of well-de-

* The *Gulf Stream* system is composed of the *Florida Current*, *Gulf Stream*, and *North Atlantic Current*. See Figure 11-6.

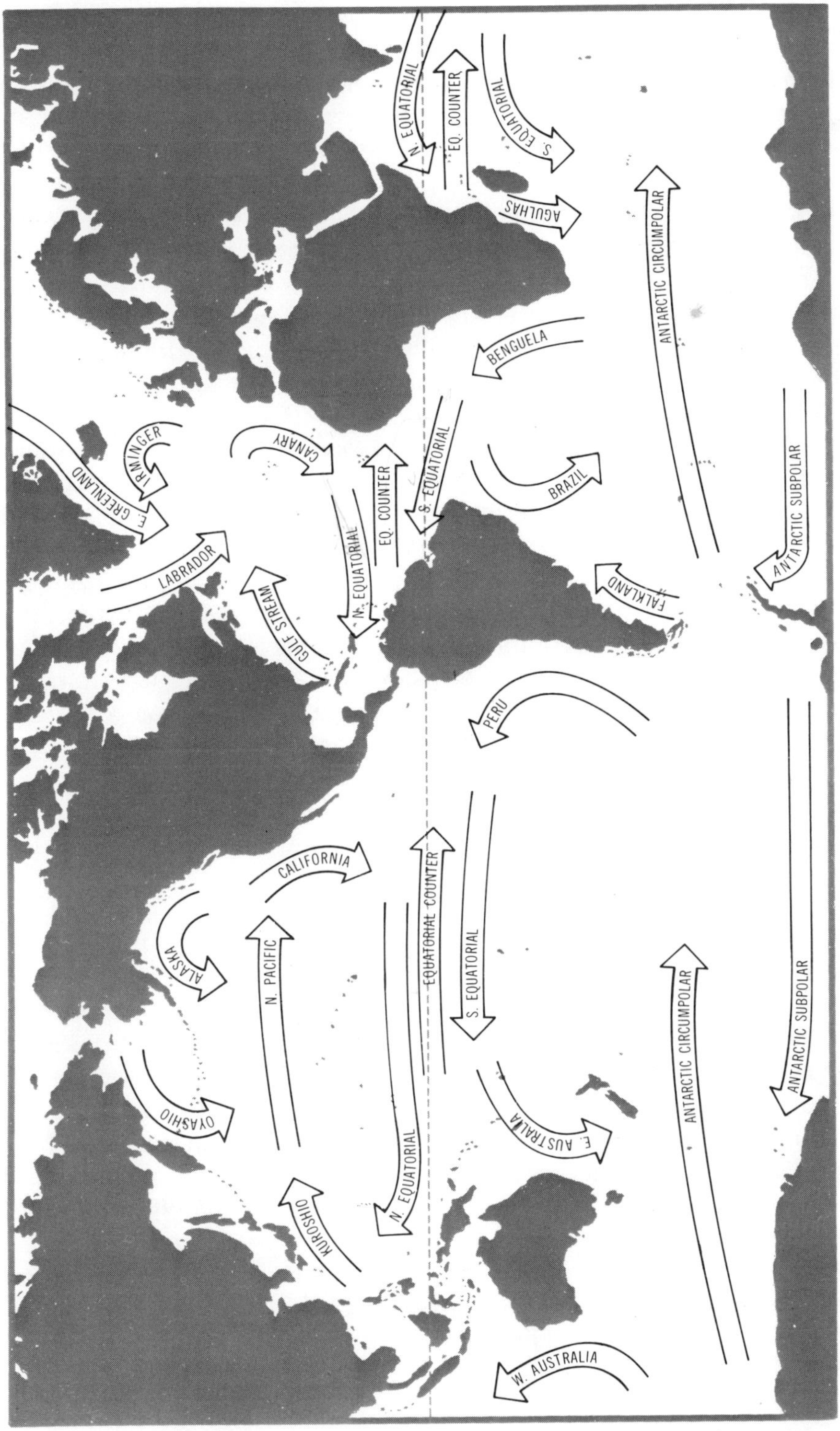

Figure 11-6 Surface Currents of the World During the Northern Hemisphere Winter

veloped subsurface currents. Two examples of these are the Pacific Undercurrent (Cromwell current) and the Atlantic Undercurrent, both of which flow from east to west within 1° of the equator. These are both well-developed currents involving transports on the order of thirty million cubic meters per second between 100 and 300 meters below the surface.

In addition the thermohaline effects have not been considered. Changes in density of surface waters produced by warming and evaporation at the lower latitudes cause a general poleward drift at the surface. This would have the effect of strengthening such currents as the Gulf Stream while weakening those tending to oppose the drift such as the Canary current in the North Atlantic. There are, of course, a number of discrepancies in the simple model. One of these is the region of the equatorial Atlantic. Here is found a large transfer of water from the South Atlantic Ocean to the North Atlantic Ocean without the separation of the equatorial gyres that appear in the Pacific. One possible explanation for this breakdown of the equatorial gyres in the Atlantic Ocean is the closeness of the African and South American land masses. Perhaps there just is not enough expanse of water to allow the gyres to develop.

Another unusual aspect of the current systems is found in the Indian Ocean. The Indian Ocean is affected by the winds resulting from the atmospheric pressure systems present over the large Eurasian continent. These monsoon winds seasonally change direction, as do the currents associated with them. Consequently, since the Indian Ocean current systems are very deeply influenced by the winds, the currents north of the equator will be toward the east in the summertime and toward the west during the wintertime.

1105 Some Representative Numbers. It might be interesting at this point to reflect on the magnitude of some of these current systems. The *Gulf Stream* is probably the most famous of all world surface currents having speeds varying from about half a knot to in excess of three knots. The amount of water transported is somewhere around 113 sverdrups* (about 30 billion gallons per second), which is more than 65 times the amount of water moved by all the rivers of the world combined. Of course all ocean currents are not of this magnitude, but even the smaller ocean currents are involved with water transports many times larger than most rivers.

Because currents necessarily involve motion, they have associated surface slopes as discussed in Chapter 9. This is certainly true for the permanent currents. In the Gulf Stream system, for example, the Sargasso

* A sverdrup (sv) is defined as a transport of one million cubic meters per second.

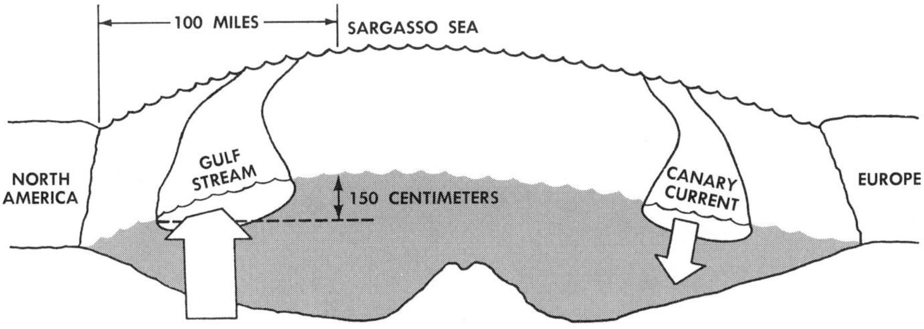

Figure 11-7 The Mound in the North Atlantic Ocean Associated with the Gulf Stream System

Sea, which is the high-pressure center of the subtropical gyre, is about 150 centimeters higher than the outside edge of the Gulf Stream itself. In other words, there is a mound of water in the center of the Atlantic Ocean corresponding to the subtropical gyre as there is in the center of all the high-pressure gyres in the world's oceans. Similarly, there are depressions in the ocean surface on the order of magnitude of 50 centimeters, corresponding to the centers of the sub-polar and equatorial gyres which are both low-pressure systems.

1106 Matthew Fontaine Maury. As indicated previously, the first man to use large amounts of ocean data in a systematic study of surface currents, from 1841 to 1853, was Matthew Fontaine Maury, a lieutenant in the U. S. Navy. Using the data accumulated from thousands of old log books, he published the first pilot charts and sailing directions for all the world oceans. As a matter of fact, pilot charts obtained today still bear the inscription, "Founded upon researches made and data collected by Lieutenant M. F. Maury, U. S. Navy."

In addition he laid the foundation for the establishment of the U. S. Weather Bureau, did most of the work in determining the location for the first transatlantic cable, and was instrumental in the establishment of the U. S. Naval Academy. He is also said to have urged the teaching of oceanography at the new institution, a piece of advice which was finally followed over one hundred years later.

1107 Applications. Ocean currents have been discussed as if they were indeed "rivers in the ocean," as Maury described them over 130 years ago. In point of fact, they may be so conceived, but if so, the bed of the river must be considered to change quite rapidly. In the Gulf Stream, for example, the path of the stream varies quite markedly from

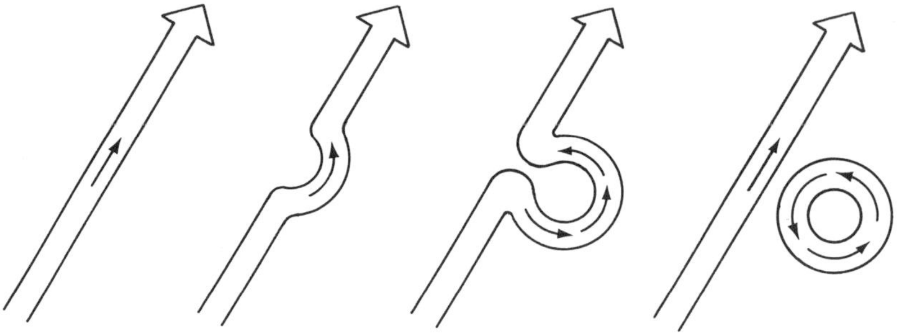

Figure 11-8 Four Stages in the Formation of an Eddy

week to week. Figure 11-8 pictures schematically the Gulf Stream on four different occasions. Note that loops form in the Gulf Stream which break off after a period of time and become eddies having associated currents which may move in a direction opposite to the stream itself.

Of course it is desirable to be able to predict the formation of these eddies, since most well-developed current systems appear to have eddies associated with them. However, at this time it is not possible to do this with the desired accuracy; about all that can be done in describing ocean currents is to indicate the average magnitude and direction of the motion at a particular location.

When one refers to a current atlas to determine average currents, typically the information is presented in the form of a *current rose* (Figure 11-9). Probabilities of current directions are shown by indicating what percentage of the time currents have been reported in what direction, and what speed they had at the time. This allows the mariner to make a good estimate of the way the water will be moving. However, it is important to realize that surface current speed and direction cannot be predicted with absolute certainty.

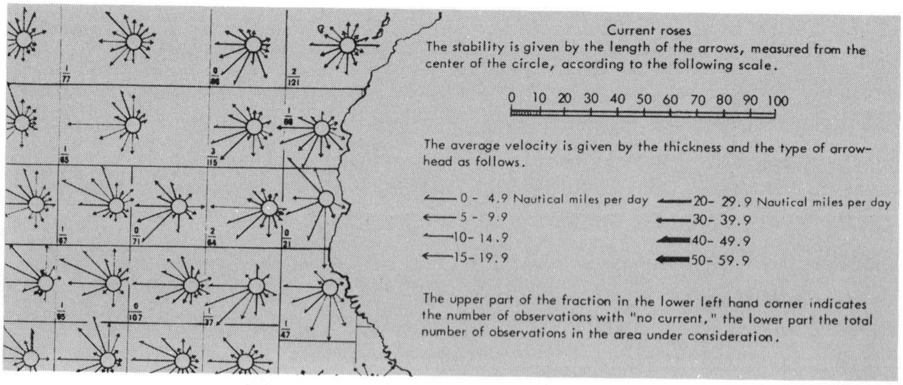

Courtesy: Koninklijk Nederlandsch Meteorologisch Institut

Figure 11-9 Typical Current Roses

Aside from the obvious effects of set and drift* on surface ships and other floating material, ocean surface currents occasionally have a fairly large effect on climate. In the main, most of the heat which is exchanged from the lower latitudes poleward is done by means of the moving atmosphere. The major exceptions to this rule are the well-developed current systems such as the *Gulf Stream* in the Atlantic and the *Kuroshio*† in the Pacific. These currents involve great volumes of water capable of carrying large amounts of heat. If these current systems pass close to land areas, and if the prevailing winds are such that warmed air is carried over these land areas, the current systems will have an effect on the climate of the adjacent land areas. An example of this is the North Atlantic Current portion of the Gulf Stream system in its passage close to the European continent. The fact that the Gulf Stream is warm and is moving into a relatively cold area, coupled with prevailing westerly winds, makes for a somewhat warmer climate in the British Isles and western Europe than normally would be expected for this latitude. This same effect occurs on the southern coast of Alaska where the effect of the Kuroshio extension (North Pacific Current) is such that this coast has a somewhat more temperate climate than would be expected.

1108 Current Measurements. Surface currents may be measured in many different ways and they have been measured for many years. Probably the easiest and most obvious way of measuring a surface current is to put a floating object in the water and observe how far and how fast it drifts. This may be a bottle, some sort of a floating drogue, a specially designed float with a radio transmitter or radar reflector for easy tracking, or even a ship itself. In actuality most current measurements which appear on pilot charts are the result of many measured ship drifts from calculated courses. If somewhat more accurate measurements are desired, various devices may be used. One of the most esoteric is the GEK (Geomagnetic electrokinetograph). The GEK consists essentially of two large electrodes which are placed in the surface water to measure the electric potential developed by a moving conductor (sea water) within the earth's magnetic field. This is basically the same principle by which a common electrical generator works, but the output is very much smaller.

One of the big problems in measuring surface currents is obtaining a measurement of water motion with respect to the earth's surface. In the deep ocean it is impossible to anchor in a manner such that a vessel does

* In navigational parlance set and drift are the direction and magnitude respectively of the current velocity vector.

† So called because the *water is very clear;* Kuroshio means *Black Current* in Japanese.

Figure 11-10 A Diver Positioning a Savonius Rotor Type Current Meter on the Bottom

not drift, so that surface currents are not accurately measured in the deep ocean, due to lack of positioning accuracy. With the advent of better navigational systems, the accuracy of current measurements at sea will improve.

However, if there is some method of fixing a current meter's position with respect to the earth, or if the drift of the device is known, rather conventional units may be used, the most common of which utilize some sort of a rotating vane. This may be a propeller, or a hemispherical cup as is used in Robinson's anemometer, or some other design of rotor, the speed of rotation being related to the current speed.

In recent years, instruments have been developed which measure rapidly fluctuating currents. This had not been possible in the past. With a rotating-vane current meter it is difficult to measure a current which changes its magnitude or direction rapidly with time. The newer devices utilize the speed of sound in two directions to determine currents; these not only take a very small period of time to make a measurement but also have no mechanical inertia. Sound-speed is measured in one direction and compared with the sound-speed measured in the opposite direction, the difference between the two being the speed of water movement.*

Most of the devices discussed here have been used with greater success in either shallow water or close to the bottom to measure bottom currents. There have been very few measurements made at sea for which great accuracies were claimed. However, a number of measurements have been made using very simple gear which have indicated the presence of currents where none had been measured before. For instance, a current cross may be lowered to the depth of interest, the angle which the line makes with the vertical is a function of the current speed. This was the case in the equatorial Pacific, for example, where the subsurface Cromwell current was first detected in 1954 by the use of this type of current meter.

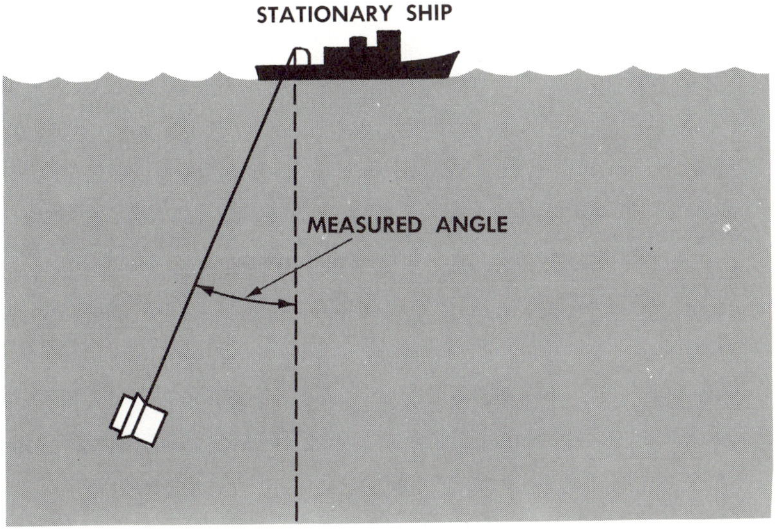

Figure 11-11 Using a Current Cross for Current Measurements

Another type of device which has been used in recent years for measuring subsurface currents is the *Swallow float*. This is a long cylinder designed to be buoyant at a particular density level, so that when it is released it will sink and remain at a particular depth. The float will then drift with the current at this level, and it may be tracked by means of acoustic gear. This has been quite successful and was utilized in affirming the previously predicted presence of a countercurrent underneath the Gulf Stream system.

* Sound energy is carried along with a moving medium.

Even though knowledge of currents at the present time is not complete, it is still sufficient for improving sailing times across the oceans. However for navigational purposes, especially in certain areas, it is many times woefully inadequate. A basic knowledge of surface currents is especially important for such obvious problems as determination of: personnel lifeboat tracks, paths of manmade pollutants, and the movements of plankton populations with their associated larger marine animals.

Additional Reading

Chapin, H., and Smith, F. G. W., *The Ocean River*, Charles Scribner's Sons, 1952.

Cotter, C. H., *The Physical Geography of the Oceans*, American Elsevier Publishing Company, Inc., 1965.

Defant, A., *Physical Oceanography*, Vol. I, Pergamon Press, 1961.

Dietrich, G., *General Oceanography*, John Wiley & Sons, 1963.

Duxbury, A. C., *The Earth and its Ocean*, Addison-Wesley Publishing Co., 1971.

Gross, M. G., *Oceanography, A View of the Earth*, Prentice-Hall, Inc., 1972.

Munk, Walter, "Ocean Currents," *Scientific American*, September 1955.

Neumann, G. and Pierson, W. J., *Principles of Physical Oceanography*, Prentice Hall, Inc., 1966.

Pickard, G. L., *Descriptive Physical Oceanography*, Pergamon Press, 1964.

von Arx, W. S., *Introduction to Physical Oceanography*, Addison-Wesley Publishing Co., 1962.

Williams, F. L., *Matthew Fontaine Maury, Scientist of the Sea*, Rutgers University Press, 1963.

CHAPTER TWELVE

Oceanic Water Masses and Their Circulation

1201 The Ubiquitous Fluids. All human life begins its existence enveloped in a mass of fluid. With birth, these babies are cast forth to spend the balance of their lives surrounded by other fluids. All human endeavors are partially, totally, or in various combinations immersed in air or water, or in the interface region of the two. These fluids may arrange themselves in large bodies of relative homogeneity called *masses*.

1202 Water Masses. A *water mass* is defined as a large homogeneous body of water which has a particular characteristic range of temperature and salinity values. The density of the water, as specified by sigma *t*, is not sufficient to identify a water mass, since a combination of various temperatures and salinities can result in the same density value.

In Chapter 7 this was demonstrated by reference to the T-S diagram with its associated sigma *t* grid. Note that since the sigma *t* curves are not straight lines, the mixing of two water masses having the same

Courtesy: Hans Marx, Baltimore, Maryland

density will result in a new mass of *greater* density. This process is known as *caballing*. For example, in Figure 12-1 water mass *a* and water mass *b* are both shown to have the same sigma *t* value. When these are mixed in equal quantities, water mass *c* results wherein $T_c=(T_a+T_b)/2$, and $S_c=(S_a+S_b)/2$, but $\sigma_{t_c} \neq (\sigma_{t_a}+\sigma_{t_b})/2$. In general, when water masses mix, resulting temperatures and salinities may be obtained by simply averaging, but resulting densities may not.

Since water masses usually gain their temperature and salinity characteristics at the surface and then seek their own density level by thermohaline convection, water masses in the ocean are categorized by two

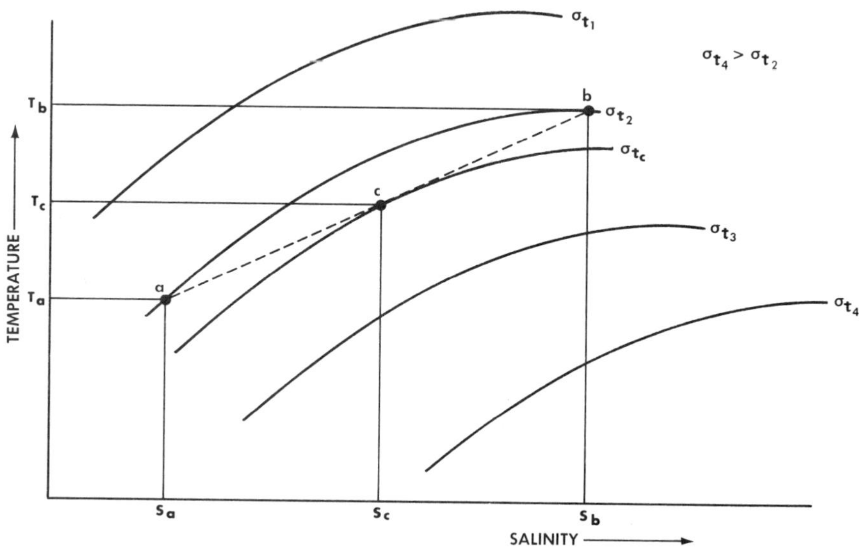

Figure 12-1 A T-S Diagram Showing Simple Mixing of Two Water Masses

factors: the depth at which they reach vertical equilibrium and the geographical source region. In order of increasing depth, water masses are classified as being *surface*, *central*, *intermediate*, *deep*, and *bottom*. *Surface* waters extend down to about 100 meters, *central* to the base of the main thermocline, *intermediate* from below the central waters to about 1,300 meters, and the *deep* and *bottom* waters fill the lower portions of the ocean basins.

The surface water is unique in that it does not fall into a true water mass category since the variability of parameters is so great.

In general, it would be expected that waters at greater depth were formed at the higher latitudes, while those existing closer to the surface were formed nearer the equator.

Table 12-1 Characteristics of Selected Water Masses

	MASS	AREA OF ORIGIN	LOCATION DEPTH (METERS)	SALINITY (°/₀₀) AND TEMP (°C) RANGE
1.	Antarctic Bottom	South Atlantic (Weddell Sea)	4,000 to bottom	34.66 (−)0.4°*
2.	Antarctic Circumpolar	South Atlantic	100–4,000	34.68–34.70 0.5°
3.	Antarctic Intermediate	South Atlantic	500–1,000	33.8 2.2°
4.	South Atlantic Central	South Atlantic	100–300	34.65–36.00 6°–18°
5.	Arctic Deep and Bottom	North Atlantic	1,300–4,000 as Deep 1,300–Bottom as Bottom	34.90–34.97 2.2°–3.5°
6.	North Atlantic Intermediate	North Atlantic	300–1,000	34.73 4°–8°
7.	North Atlantic Central	North Atlantic	100–500	35.10–36.70 8°–19°
8.	European Mediterranean	European Mediterranean	1,400–1,600	37.75 13°
9.	Pacific Equatorial	Central Pacific	200–1,000	34.60–35.15 8°–15°
10.	Indian Central	Indian	100–500	34.60–35.50 8°–15°
11.	Red Sea	Red Sea	2,900–3,100	40.00–41.00 18°
12.	Black Sea	Black Sea	0–200	16.00 (Average) Various Temp.

* This is the only negative temperature in this table.

1203 Atlantic Ocean. In the immediate vicinity of the Antarctic Continent, particularly the Weddell Sea, waters reach extremely low temperatures in the winter. Due to this low temperature and high salinity resulting from ice formation (see Chapter 13), this water has the highest sigma *t* of any in the world ocean. As a consequence, having once gained these characteristics, it sinks and flows along the ocean floor in a direction toward the equator. In fact, traces of this water have been measured as far as 45° *North* latitude. This water mass is called *Antarctic Bottom Water*, obviously because of its location and formation area. The Antarctic Bottom water mass also flows eastward around the Antarctic Continent due to the surprisingly deep-reaching effects of the surface West Wind Drift, mixes well below the surface with masses on its north edge, and becomes a separate, fairly homogeneous mass known as *Antarctic Circumpolar Water*. The deeper reaches of this mass, as it flows eastward, continuously provide deep water to the Indian and South Pacific Oceans. While it is true that some water circumnavigates the continent, it has been difficult to estimate the amount.

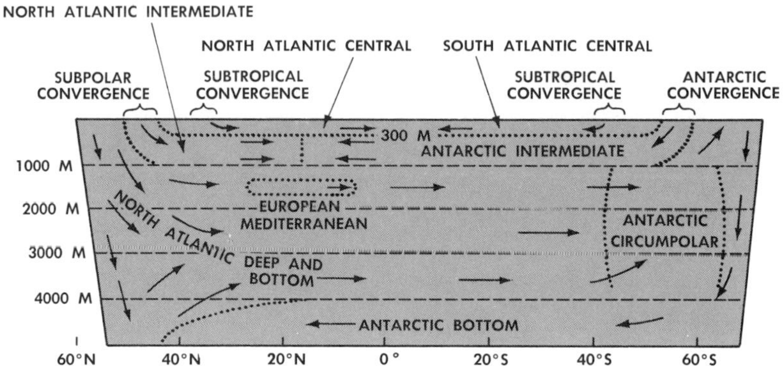

Figure 12-2 Atlantic Ocean: General Subsurface Movement

The *Arctic Deep and Bottom Water* (North Atlantic Deep and Bottom Water) is formed in relatively small areas off the coast of Greenland, one of which is the convergent region produced by the Irminger and East Greenland currents.* North Atlantic Deep and Bottom Water, less dense than Antarctic Bottom Water overrides Antarctic Bottom Water all the way to the South Atlantic reaching the surface south of 60°S (see Figure 12-2). The North Atlantic Deep Water is continuously modified in its transit by mixing with masses yet to be discussed.

The *Antarctic Convergence Zone*, located at approximately 60°S latitude, is primarily produced by the seasonal cooling of the Antarctic Intermediate Water as it sinks to its density level. This particular convergence zone is present at nearly all longitudes of the earth; however, similar convergence zones in the North Atlantic and North Pacific are somewhat discontinuous and at times can be difficult to locate. North Atlantic Intermediate Water flows south from the Arctic Convergence to approximately 20°N where it mixes with Antarctic Intermediate Water.

North and South Atlantic Central Waters form at the surface at their respective subtropical convergences during the winters. They sink and flow toward the equator losing their identities as they spread.

* Periodic overflows from the North Polar Sea across the Greenland-Scotland Ridge cascade down the southern slope with relatively high velocities due to the water's very cold (−1.4°C) temperature (the coldest water anywhere in the deep sea); it is less dense, however, than North Atlantic Deep and Bottom Water because of a lower salt content.

The one significant incursion of foreign waters is the large mass of European Mediterranean Water which finds its level at the average depth of 1,500 meters, after leaving through the Strait of Gibraltar. This water mass is continually formed in the northern area of the western Mediterranean by winter cooling and evaporation by the dry air sweeping north

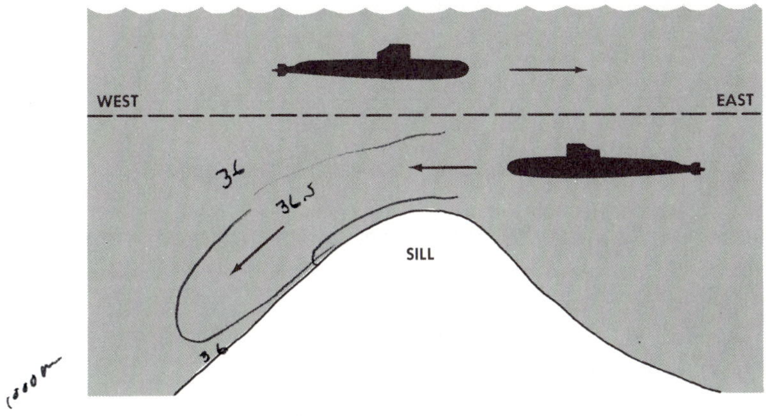

Figure 12-3 Water Flow in the Strait of Gibraltar

from Africa. The cool, saline water sinks, flows south and west, and then spills out over the sill. On the surface in the Strait the less dense Atlantic waters flow in to maintain the balance, creating a two-layered stratification with each layer flowing in opposite directions.

During World War II German submarines are said to have used the flows to transit the Strait undetected. They would dive deep or shallow depending on whether they desired to exit or enter, compensate for the required neutral buoyancy state, and then ride the flow quietly without use of their motors. This was a very ingenious use of environmental knowledge to circumvent detection.

The Mediterranean Water, with its increased salinity, has strong effects on the upper section of the North Atlantic Deep Water mass. Although its influence is felt to the west and south predominately, its telltale salinity maximum has been traced to locations up to 1,500 miles from Gibraltar.

In conclusion, the Atlantic Ocean is constantly renewing itself at all depths although at a very slow rate. Recent analyses utilizing radioactive carbon measurements indicate that it has been about 750 years since Antarctic Bottom Water in the Atlantic was at the surface. In contrast to this, 1,500 years is estimated for the age of this water mass in the Pacific Ocean.

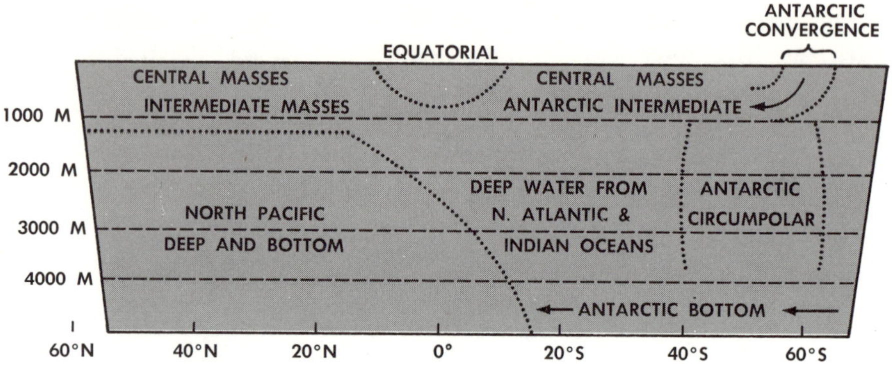

Figure 12-4 Pacific Ocean: General Subsurface Structure

1204 Pacific Ocean. The Pacific Ocean is noted for its generally sluggish deep-water flow pattern when compared to the other oceans. However, the Antarctic Bottom Water, as it flows around the Antarctic Continent, provides a fairly continuous input to the South Pacific Basin. On the other hand, Antarctic Circumpolar Water, which has been partially mixed with waters of the Atlantic and Indian Oceans, enters from the west to slowly but continuously push into the South Pacific deep layer (below 1,000 m).

The intermediate and central layers of the entire Pacific Ocean are diffuse and not well defined. The various convergence zones which would be analogous to those in the Atlantic are discontinuous and misplaced. Several masses existing at the same depth in different areas make a cross-sectional depiction difficult to construct. It is important to note that there is near-surface mixing of masses from distant regions at the equator forming the Pacific Equatorial Water Mass—the only major water mass which does not receive any characteristics from the surface near the formation area.

The North Pacific is unique because no extremely dense water masses form in its most northerly reaches. The Deep and Bottom Water of the North Pacific experiences little interchange with other areas. Its origin is in doubt both in time and space, and it is characterized by an oxygen minimum due to the sluggish flow present.

Because of the slow movement of the subsurface mass, the surface current motion reaches deeper and has greater effect on the subsurface characteristics in the North Pacific than do the surface currents in other oceans. This is probably due to the general absence of vital thermohaline convective activity in the North Pacific Ocean.

1205 Indian Ocean. Of the three major oceans only the Indian Ocean does not extend into the Northern Hemisphere. There is no cold

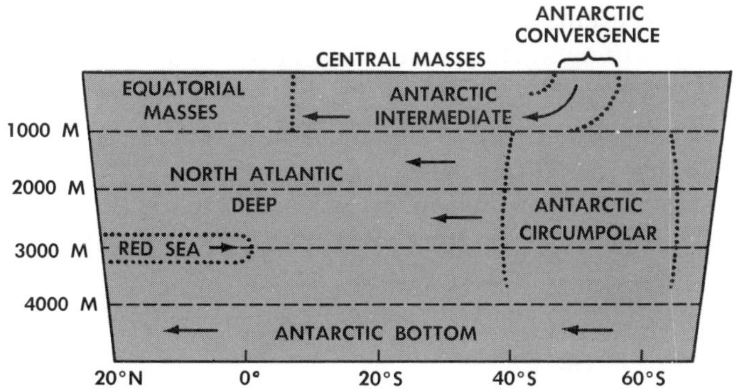

Figure 12-5 Indian Ocean: General Subsurface Structure

water sinking along its northern edge causing the deep water mass to have a lesser movement than that in the North Atlantic Ocean. However, there is a well-defined bottom flow in the South and, oceanographically speaking, it is like the South Atlantic south of the Subtropical Convergence Zone at about 40°S latitude.

The Antarctic Bottom Water is present at all latitudes of the Indian Ocean. The deep layer is that which is led around the south tip of Africa from the Atlantic; it is reasonably well oxygenated, especially considering the distance from its source region in the North Atlantic. The Antarctic Intermediate Water forms at the Antarctic Convergence Zone and spreads to the north. The Indian Central Water sinks at the Subtropical Convergence and flows north toward equatorial regions.

Bottom water from the Red Sea flows over the sill and on through the entrance at the Strait of Bab el Mandeb to spread and mix with deep layers of the Indian Ocean. Red Sea Water is characterized by its very high salinity of about 40‰ to 41‰. This water mass is formed within the Sea by constant evaporation by dry air from Africa and by winter cooling periods in much the same manner as European Mediterranean Water. High salinity causes its density to be such that it spreads out in the Indian Ocean at depths near 3,000 meters. Traces have been identified as far as 1,250 miles south of the Gulf of Aden. Red Sea Water provides the only significant modifying effect in the entire deep Indian Ocean. As an aside, hot spots have been discovered recently at great depths (2,040 m) in the Red Sea. The anomalous temperatures measured thus far range from 22°C to 56°C; salinities have been determined to be in excess of 250‰. Their causes remain unexplained but future concerted investigations are planned to develop answers as to how the spots have been formed and how they continue to exist.

The equatorial shallow layers of the Indian Ocean are not clearly defined. This is partially due to seasonal monsoon changes of surface currents. The water is being constantly overturned by changing winds and does not have significant characteristics. Little distinguishable subsurface flow is present.

1206 Black Sea—A Sea Apart. The Black Sea, with its complete lack of thermohaline convection, has a complete oxygen disappearance at all depths below 200 meters. Precipitation and runoff far exceed evaporation. The flow into the sea through the Bosporus from the European Mediterranean is so meager that it would renew the waters below 30 meters only once in 500 years. Consequently, the deep waters have become stagnant; hydrogen sulfide is present; and only anaerobic bacteria can live in the blackened waters.

1207 Conclusion. The preceding discussion, at best, is only a very cursory qualitative treatment of deep-water ocean circulation. Although the mechanisms producing subsurface flows were discussed, bottom topographic effects have been neglected. But to ignore the latter in a detailed study would be a serious omission. Only the most prominent of the marginal sea effects have been introduced.

It will be seen in later chapters how the air can be subdivided into masses displaying identifiable characteristics similar to those of water masses of the oceans. In addition, there can be identified clear boundaries between these masses, both in the sea and in the air. Those in the sea have not achieved the importance that those in the atmosphere enjoy; however, the wall of the Gulf Stream has been clearly identified as a "front" in the ocean. The boundaries or "fronts" between masses in the atmosphere have become essential to the analysis of weather and will be discussed in considerable detail later.

Additional Reading

Bailey, H. S., Jr., "The Voyage of the 'Challenger'," *Scientific American*, May, 1953.

Dietrich, Gunter, *General Oceanography*, John Wiley and Sons, New York, 1963.

Duxbury, A. C., *The Earth and Its Ocean*, Addison-Wesley Publishing Co., 1971.

Gross, M. G., *Oceanography, A View of the Earth*, Prentice-Hall, Inc., 1972.

King, Cuchlaine A. M., *An Introduction to Oceanography*, McGraw-Hill, Inc., 1963.

Kort, V. G., "The Antarctic Ocean," *Scientific American*, September, 1962.

Munk, W., "The Circulation of the Oceans," *Scientific American*, September, 1955.

Stommel, Henry, "The Anatomy of the Atlantic," *Scientific American*, January, 1955.

Sverdrup, H. V., Johnson, Martin, W., and Fleming, Richard, H., *The Oceans*, Prentice-Hall, Inc., 1942.

Weyl, P. K., *Oceanography*, John Wiley & Sons, New York, 1970.

Williams, Jerome, *Oceanography*, Little, Brown and Company, Inc., Boston, 1962.

Yasso, Warren, E., *Oceanography*, Holt, Rinehart and Winston, Inc., New York, 1965.

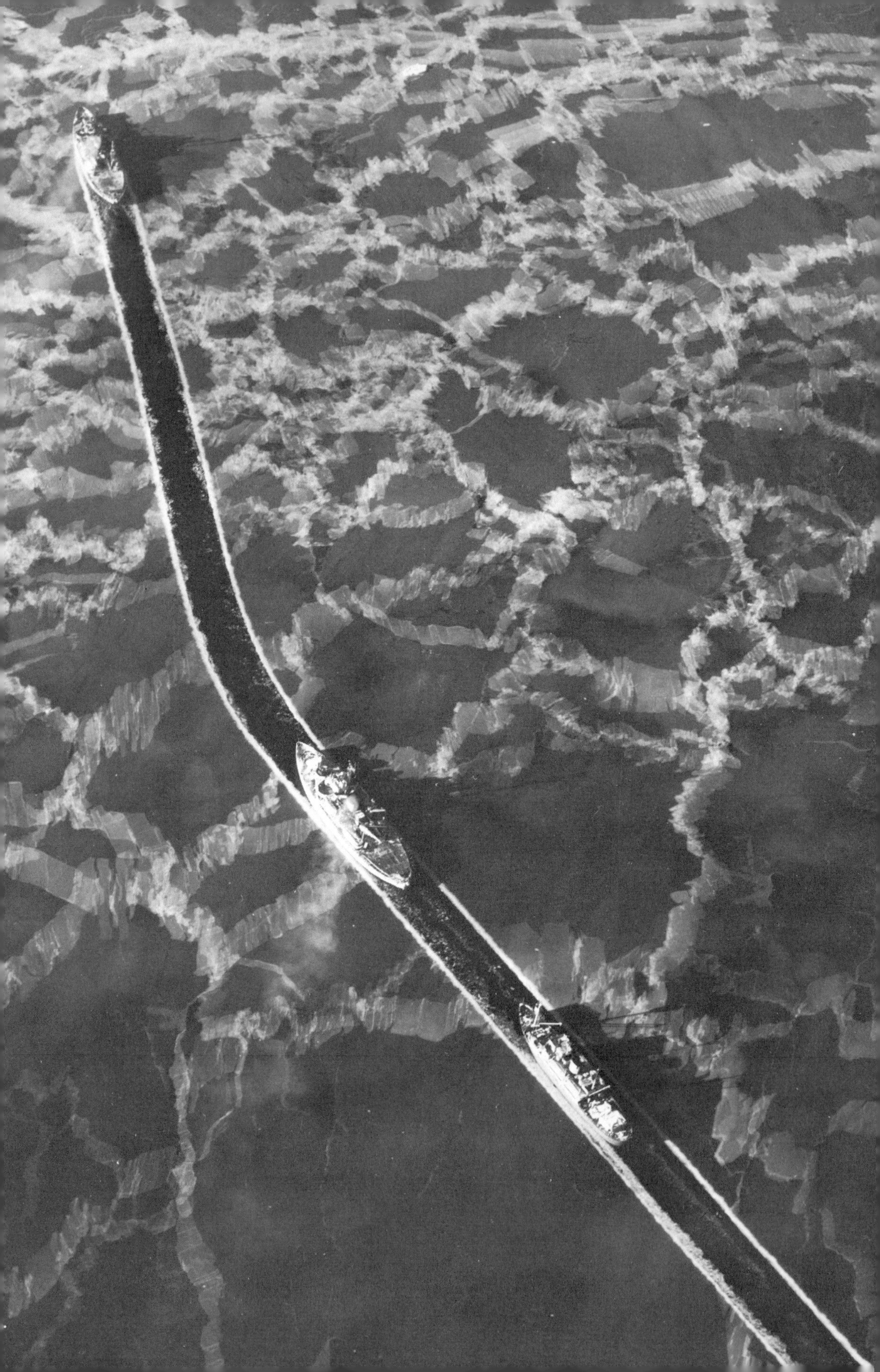

CHAPTER THIRTEEN

Ice: Formation and Movement

1301 Introduction. Ice at sea has been a hazard to navigation ever since sailors began taking their ships to the colder reaches of the world. Land ice (icebergs) can and have cut great gaping holes in ships causing loss of life and property. Sea ice has trapped ships for long periods, sometimes crushing them in the pressure vise tightened by wind and currents. Each type of ice forms and disintegrates quite differently.

1302 Ice Formation. When fresh water cools, it increases in density until its maximum density temperature of +4.0°C is reached. As further cooling takes place, the surface layer continually becomes less dense until freezing occurs at 0°C. As salts are added to pure water, the temperature of maximum density decreases linearly as does the freezing point, in accordance with the graph shown in Figure 13-1. The maximum density and freezing point curves coincide at 29.61°F (−1.33°C) at a salinity of 24.7 parts per thousand. The freezing point continues its linear decrease

Icebreaker USCGC Edisto *cracks the pack ice off Palmer Gerlasche Strait for USCGC* Northwind *and RRS* John Biscoe for all salinities above 24.7 parts per thousand, so that the water freezes before the temperature of maximum density is reached for all salinities above this value.

Fresh water as it is cooled to 4°C at the surface continually sinks and, in the process, builds a column which has the 4°C water at the bottom in a completely stable condition. This sinking continues to progressively lesser depths until the entire column is at 4°C. After this condition is reached, there can be no more convection due to temperature change. Further cooling at the surface takes place rapidly until the freezing temperature (0°) is reached and a thin sheet of ice is formed.

The thin sheet begins to act as an insulator progressively slowing the ice's thickening rate because the water beneath is above freezing. Beyond this stage, water temperature can be reduced only by conduction.

A contrary situation occurs in the salt water body. In this case, the most dense water is colder than the surface layers. The entire water column must be cooled to the freezing point before ice can form because the surface water continually sinks as it is cooled. After the entire column is at the freezing temperature, latent heat is removed by both radiation and conduction to the overlying air. The ice sheet then forms.

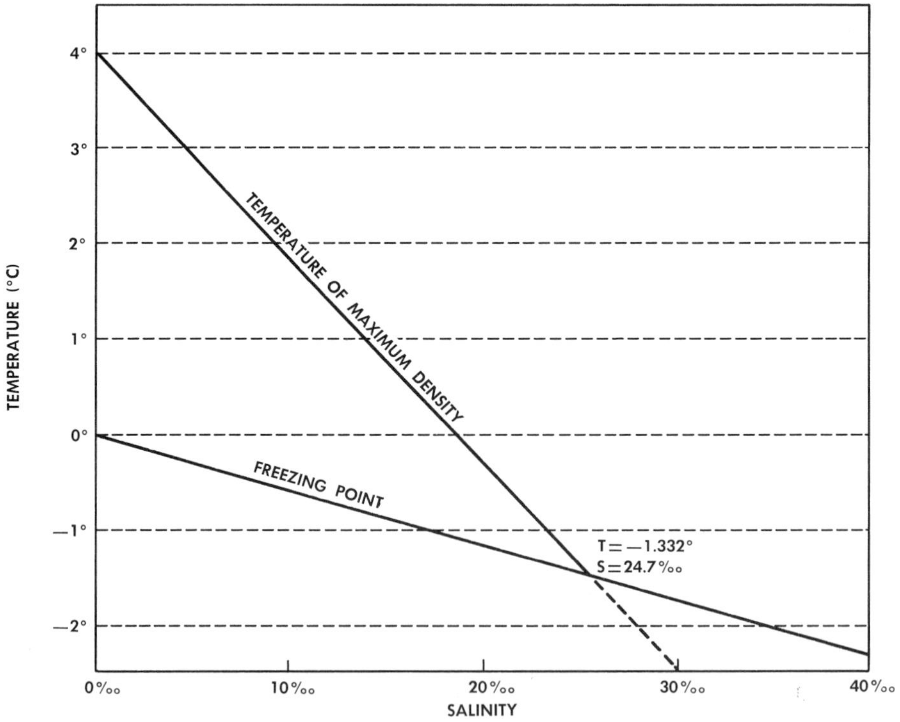

Figure 13-1 Relationship Between Temperature of Maximum Density and Freezing Point for Water of Varying Salinity

The result is that the fresh water body with the same depth as a given salt water body will be covered with ice in a significantly shorter period. However, as the ice layers on each are formed, the salt water ice will increase in thickness faster. Less heat has to be removed from salt water directly beneath its ice than must be removed from fresh water to continue the ice formation.

Interestingly, if it were possible to produce an artificial heat source on the bottom of the Great Lakes, maintaining a bottom water temperature of 5°C, a continuous upwelling of water with a lesser density than that of the colder surface water would be produced during the winter. The warmer water then would moderate those cold polar outbreaks from Canada allowing a general lessening of climatic weather severity for those states south and east of the Lakes. Also this possibly would allow the lakes' shipping waters to be navigable the entire year with the attendant economic and military benefits.

1303 Sea Ice. Sea ice forms on the surface of sea water by progressive freezing. A greasy or oily appearance will be the first indication with individual particles or spicules soon becoming visible. The surface becomes slushy as the particles increase in number causing a noticeable reduction in wave height. The particles freeze together in a thin sheet with extreme plasticity permitting the ice to conform to any shape required by the forces acting upon it. For instance, pancake ice forms when small cakes, chunks formed from slush, bumping against each other, raise the edges of each piece. As the sea ice thickens and becomes larger in area, the ice floe is formed. Floe sizes range from 10 meters up to 5 nautical miles in major dimension. Larger solid ice sections are called ice fields.

Ice, up to the age of one year, is called *winter ice*. As it ages, it grows harder and more brittle. Since only the pure water can freeze, brine bubbles form, thereby decreasing the salt content of the frozen volume to as low as 2 parts per thousand in winter ice. The ice becomes leached as the brine seeps down, causing the water just beneath the ice to become more saline. *Melt water* puddles on sea ice one year or older, if not contaminated by spray, are generally fresh enough to replenish the potable water supply of a ship. Older ice is generally even more salt-free.

As ice thickens, it provides increased insulation and slows the rate of heat transfer from the water beneath. Snow will protect the ice from the intense cold above also slowing the heat transfer. Both of these factors inhibit the freezing rate of the water beneath the ice. Winter ice can grow to a maximum of six feet in thickness. *Polar ice*, as sea ice is known after the age of one year, has been measured in thickness up to fifteen feet. Frozen snow, after compaction, is an important factor in causing these extreme thicknesses. These thickness measurements are representative only of smooth, unbroken ice, since actual thicknesses may be increased by a factor of two or three by rafting, as discussed later.

The density of ice is also dependent upon its age. Pure frozen water has a specific gravity of 0.917. Newly formed sea ice averages 0.925 g/cm^3 in density, which decreases with aging to about 0.85 g/cm^3. The latter minimum value exists because of air entrapment, the voids produced when the brine leaches out.

Sea ice can take many forms when affected by its environment. When floes encounter each other with force, one may override the other, a condition known as *rafting*, or each edge may rise above its original level, commonly called *tenting*. Sea ice with irregular topographic features caused by squeezing, is called *pressure ice;* a line of such ice is said to be a *hummock* (pressure ridge). A hummock can protrude simultaneously up and down to an average of forty feet in each direction. Extreme total thicknesses of 150 feet have been observed. A submarine, by using its sonar, may be continually aware of these hummocks which is especially helpful when searching for an open or thin spot to surface.

Nearshore ice ridges, which are results of strong currents, prevailing winds, and local bottom topography, have been observed to have thicknesses of 60 to 70 feet with extremes to 200 feet.

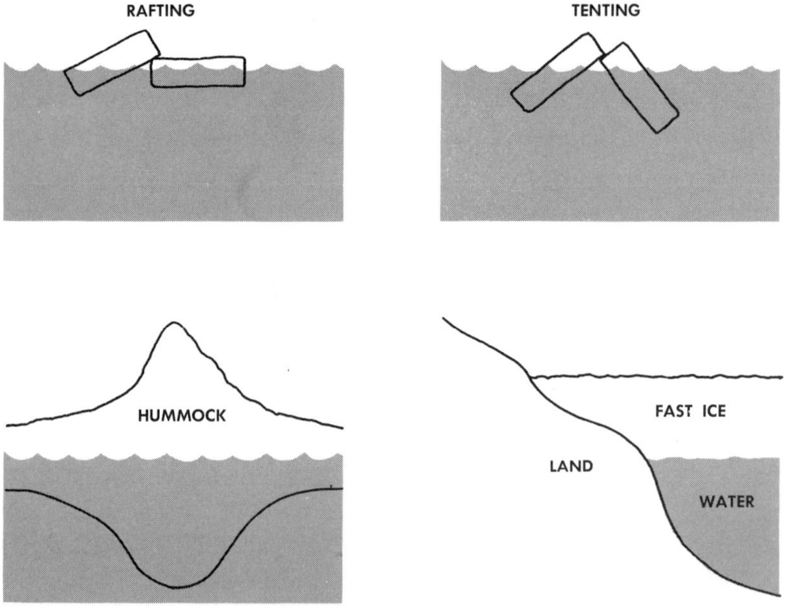

Figure 13-2 Ice Forms

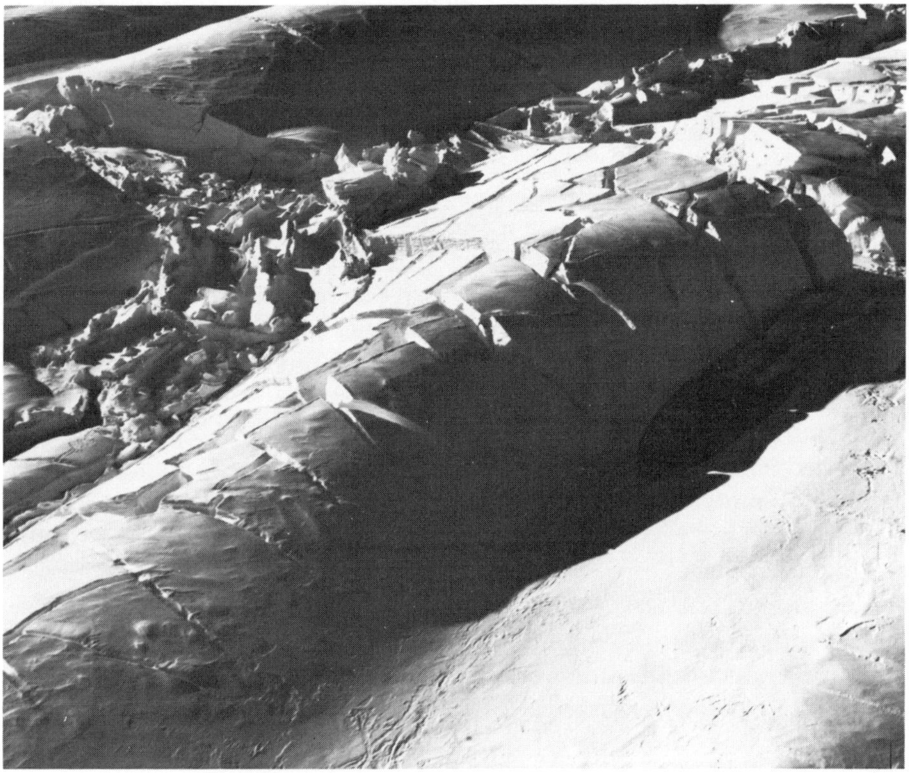

Figure 13-3 Hummock in Bay of Whales, Antarctica

Openings in sea ice can result from the changing environment. A long narrow crack large enough for a ship's passage is called a *lead*. Helicopters are often used to patrol in the direction of desired travel; their reports assist the commander in deciding which lead to take. Sometimes the leads close in such a manner to leave large enclosed bodies of water called *polynyas* (Russian, *polye*—field). A recurring polynya, maintained by upwelling warm water, occurs in a northerly section of Baffin Bay.

The main area of the North Polar Sea is covered with pack ice including many floes, ridges, hummocks, and leads. There are also ice islands which the United States and the Soviet Union have manned from time to time to gather weather data and conduct other scientific projects. The islands are separated sections of the shelf ice (tabular icebergs formed in the continental shelf area) many years old. Long-term tracking of these ice islands has evidenced a prevailing surface flow pattern in the clockwise direction.

1304 Disintegration. The melting and disintegration process of a large ice floe generally follows a sequence of events, each contributing in its own way to the breakup. The probable snow cover must first melt to allow a faster downward heat flux through the ice. Water puddles form from the melting ice and snow lowering the albedo and therefore allowing easier absorption of the sun's insolation. As the melt water drains off, the floe becomes more buoyant and may start cracking due to the newly created stresses caused by changing positions. In addition, the winds, tides, and currents will assist the floe in grinding itself into other floes or the sea bottom if in shallow water. Dirt particles and algae may have become embedded in the floe in its later life which would further decrease the albedo (by darkening the color) causing an additional heat absorption, which increases the temperature and the melting rate. Possible advection of warmer water to the floe or vice versa also will play a contributing part.

If the surface of the floe has become honeycombed due to the brine solution having settled out and initial melting, the remainder is called *rotten ice*. This process occurs over a very short period after initial spring warming.

The maximum ice coverage* usually occurs during late March or early April; the Navy Weather Service issues general bulletins on sea-ice conditions during the summer for all affected areas and, on request, for any specific area.

1305 Northern Hemisphere Icebergs. Icebergs represent an entirely different hazard to shipping than does sea ice. They are separate blocks of *land ice* which can be intact far from their points of origin. They are produced from land glaciers in Spitzbergen, Novaya Zemlya, Alaska, and Greenland, among other lands bordering the northern seas. Nearly all the icebergs *calved* (broken away from parent glacier) in places other than Greenland melt near their birthplaces. However, Greenland produces so many each year that many bergs find their ways to the trans-Atlantic shipping lanes to create hazardous conditions in spring and summer. This prolific production is the result of the glacier movement over the ground at speeds up to 100 feet per day.

Some icebergs break off from the East Coast of Greenland and follow the East Greenland Current down and around Cape Farewell. They drift in the Davis Strait and then are taken by the Labrador Current down to the Grand Banks. The large majority, however, come down from the West Coast of Greenland proceeding through the Davis Strait. The process, from calving to appearance in shipping lanes, may take two winters for a single berg—especially if calved on the East Coast.

* See H.O. Pub. No. 550, *Ice Atlas of the Northern Hemisphere* for a more complete depiction of ice coverage.

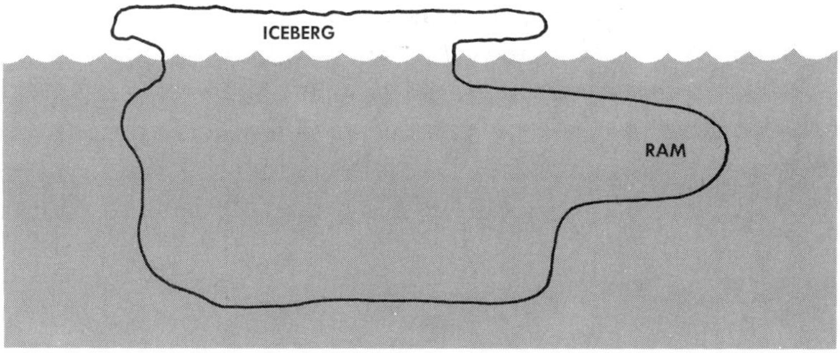

Figure 13-4 Iceberg Profile

The irregularly shaped icebergs of the Northern Hemisphere are not large when compared to the great tabular icebergs of Antarctica. The largest Northern Hemisphere iceberg ever recorded—seven miles by three and a half miles—was sighted off Baffin Island in 1882. However, they are generally much smaller. A berg the size of a house is called a *bergy bit;* one somewhat smaller which can still inflict ship damage is called a *growler*—this due to its noise as it bobs in the water. Also a principal danger which an iceberg may have is the underwater protrusion called a *ram*.

Icebergs, being land ice, have a complete fresh water composition except for that small amount of *morainal* glacial material embedded within the ice. One may consider that the berg has a density of 0.920 gm/cm^3. If the berg is floating in the sea with a water density 1.028 gm/cm^3 (salinity content of 35 parts per thousand at a temperature of 30°F), it will have 87.5 percent of its mass below the surface and, if uniform in shape, will have a depth extension seven times the height above the sea surface.

1306 Southern Hemisphere Icebergs. The *tabular icebergs* of Antarctic origin are quite different in nature. They come from the great shelf ice present in many coastal areas. The uniform pieces break off the outer edges with dimensions measured in miles. The largest iceberg ever reported was sighted by the USCGC *Glacier*, about 150 miles west of Scott Island (178°E). It was 60 miles wide and 208 miles long, larger than twice the size of Connecticut. That iceberg was 100 feet tall, but some rise more than 300 feet in the air which means that they extend more than half a mile below the sea surface.

1307 Ice Movement. A large number of icebergs within a small area indicate that the bergs are influenced by the prevailing currents as well as the more obvious winds. This is to be expected since the greater portion of the berg is below the water surface and therefore would tend to move with the water as described in Chapter 9. When the icebergs follow a large-scale continuous current such as the Labrador Current, the trajectory path will be relatively unaffected by the winds.

Sea-ice movement is predictable when a prevailing wind is the cause. Because of the wind-induced surface-water effects discussed in the Ekman Spiral Theory and the frictional effects of water on the surface winds, the sea ice moves parallel to the isobars or nearly so. (See Chapter 9.) In the Northern Hemisphere the atmospheric low is to the left of the ice motion; thc icc moves at a speed one to seven percent that of the wind.

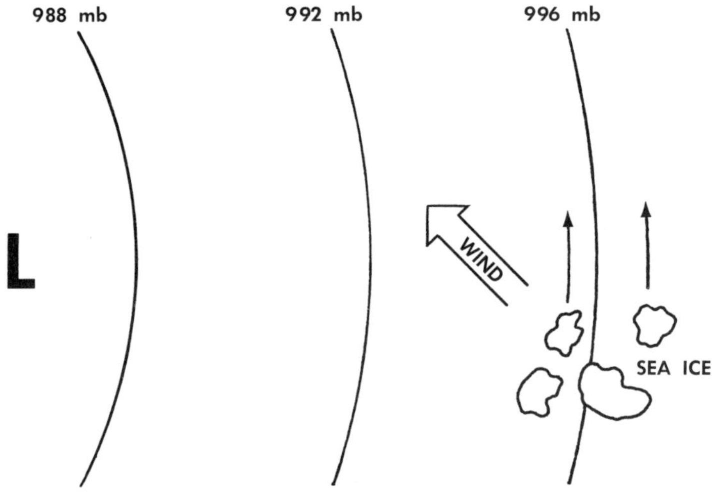

Figure 13-5 Sea-Ice Movement in the Northern Hemisphere

These observations have been verified repeatedly and found to be compatible with computerized numerical models. As a result, successful ice-field forecasting is now one of the services available to polar naval operations.

1308 Maritime Ice Operations. For operations in ice areas it is strongly advisable to have ships with reinforced hulls as well as genuine ice-breakers. The crews must be trained and well equipped. A sharp lookout is always maintained and all bergs given a wide berth. When in pack ice the primary hull area to be kept clear is the stern, because of the vulnerability of rudders and screws.

Figure 13-6 Iceberg Movement Caused by Subsurface Current through Broken Sea Ice

Radar is a valuable aid for iceberg detection—although not infallible—but visual detection remains the best method of gaining hints of sea-ice presence. "Ice blink," or yellowish glare indicating sea ice, can be in the sky or on the horizon; "water sky," or dark tint, means clear water ahead; and a white reflection indicates snow or ice on land nearby. A sharp drop in injection temperatures* and the gradual lessening of swell are also observable hints of pack-ice presence.

The International Ice Patrol was organized by treaty in 1913 as a direct result of the *Titanic*'s sinking, in an effort to maintain close observation of all icebergs in the North Atlantic trade routes. The U. S. Coast Guard keeps ice-reconnaissance ships and aircraft active throughout the spring and early summer ice season. The Coast Guard provides data through the forecasting centers and issues ice sightings directly on request to ships in the area. Merchant ships also transmit data pertinent to the effort of maintaining route safety.

1309 Conclusions. This chapter described the basic important ice formations with which all sailors should be familiar. The *ice cycle* is one

* Temperature of water drawn into ship's main engine cooling system.

of the fundamental air-ocean interactions. Too little is known because environmental conditions (it gets cold and rough!) have driven the bulk of otherwise interested scientists to more temperate climates to conduct basic research. The fact remains, that sea and land ice are omnipresent hazards to polar operations.

Additional Reading

Bowditch, N., "Oceanography," Part Six of *American Practical Navigator* (1958), U. S. Navy Oceanographic Office, 1962.

Boyle, R. J., *Ice Glossary*, U. S. Navy Electronics Laboratory, San Diego, 1965.

Dietrich, Gunter, *General Oceanography*, John Wiley and Sons, New York, 1963.

Duxbury, A. C., *The Earth and Its Ocean*, Addison-Wesley Publishing Co., 1971.

Gilluly, J., Waters, A. C., and Woodford, A. O., *Principles of Geology*, 2nd ed., W. H. Freeman and Company, 1959.

H. O. Publication No. 606-d, *Ice Observations*, revised edition, 1966.

Kort, V. G., "The Antarctic Ocean," *Scientific American*, September, 1962.

Pounder, Elton R., *Physics of Ice*, Pergamon Press, New York, 1965.

Robin, G. deQ., "The Ice of the Antarctic," *Scientific American*, September, 1962.

Shepard, Francis P., *Submarine Geology*, 2nd ed., Harper and Row, Inc., New York, 1963.

Sverdrup, H. V., Johnson, Martin W., and Fleming, Richard H., *The Oceans*, Prentice-Hall, Inc., 1942.

Williams, Jerome, *Oceanography*, Little, Brown and Company, Inc., Boston, 1962.

CHAPTER FOURTEEN

Life in the Sea

1401 Introduction. *Marine biology* is the common term used to describe the study of life in the sea. A strict connotation of the term is that it is the detailed study of the individual, be it plant or animal. However, the requirements of the broader science, *biological oceanography*, now necessarily include studies of the physics, dynamics, biochemistry, and meteorology of the oceans. With this knowledge the scientist is properly equipped to approach the problems of fisheries, estuarine ecology, as well as biological effects on antisubmarine warfare. Questions such as "Where will the cod be next year?" are now being reasonably answered. A Chesapeake Bay study group headed by West Coast physical oceanographers and formed to study the water body pollution problem discovered that living individuals could not be treated as things "simply present." Their bodily functions, habits, and chemical composition had to be understood so that the coordinated effects could be comprehended. In short, the problems shall be solved only if an integrated discipline is followed.

The bathyscaph Trieste *in Coronado Submarine Canyon at 580 fathoms*

1402 Oceanic Zones. The ocean areas have been divided and classified into several convenient zones to allow discussion of the organisms by location and function as that particular aspect is directly affected by the individual's habitat.

The classifications of marine environments (Figure 14-1) are presented to assist in understanding the various areas. Note the two basic divisions, *pelagic* (entire water mass) and *benthic* (bottom). The first division includes the provinces of the *neritic* or shallow-water environment along the seacoast and the *oceanic* or deep-water marine (open sea) environment. The second division, the benthic (Greek, *benthos:* depth of the sea) is composed of the tidal zones, the bathyal zone, the abyssal zone, and the hadal zone. The detailed breakdown of the two major divisions is explained in the figure.

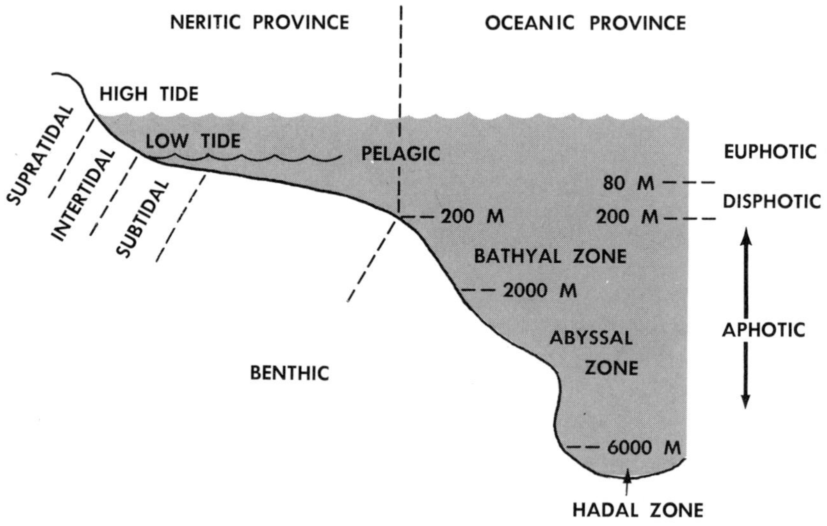

Figure 14-1 Classifications of Marine Environments

For convenience, the neritic and oceanic provinces are arbitrarily divided at the *continental shelf break*, which is approximately the same as the 200-meter depth universally accepted as the effective limit of light penetration. The point which must be made from the dual consideration is that in regions where the shelf break is much more shallow than 200 meters, that depth is reached in a very short distance seaward of the shelf break. This maintains the basic integrity of the various definitions.

The nearshore benthic division is divided according to the tidal fluctuations. The *subtidal zone* as shown is arbitrarily cut at its seaward edge where the *spring tides* (Chapter 20) reach their lowest point. On many coasts this is a considerable distance from the mean low-water point or seaward edge of the intertidal zone. The benthic life is greatly influenced by the water presence in this general area.

The *euphotic* (Greek, well lighted) or light zone of the pelagic division is the light-affected volume. Photosynthesis is limited to this depth. Of course the actual depth of this zone fluctuates with turbidity and seasons, but in the deep ocean it appears to average about 80 meters. The next light layer below the surface is known as the *disphotic zone* (Greek, poorly lighted) (80 m to 200 m); light penetrates this zone in significant intensity but not enough to permit active photosynthetic activity. Below 200 meters is the *aphotic zone* (Greek, lightless), that region where only biologically insignificant light normally penetrates.

The *abyssal* and *hadal* zones were considered to be without life until the last hundred years. The *Trieste* descents were the final proof that life does in fact exist—however sparse and difficult—at the greatest depths.

1403 Dissolved Gases. Several gases are being constantly dissolved into the ocean's surface layers partially because they are in direct contact with the atmosphere. The most common of these are oxygen, carbon dioxide, nitrogen, argon, helium, and neon. The last four are chemically inert. Hydrogen sulfide is also present in stagnant, deoxygenated water bodies (Chapter 12). Being non-atmospheric in origin, the gas is derived from biological or decaying chemical actions.

Oxygen is dissolved directly from the atmosphere at the air-ocean interface or is chemically released by photosynthesis in the water (see following section). Wind-driven turbulence in the upper layers aids in introducing oxygen to lower layers; this process coupled with thermohaline convection maintains the world's oceans in a predominately non-stagnant state. The presence of oxygen is, of course, necessary for marine animal life and it assists the physical oceanographer in tracing, identifying, and labeling water masses.

Carbon dioxide, the necessary gas for photosynthesis, is replenished by animal respiration and by dissolving action directly from the atmosphere. Experimentation is presently in progress to determine the importance of this latter method.

1404 Phytoplankton. *Plankton* (Greek, *planktos*—drifter) describes the free-floating plant and animal life in the ocean. The planktonic organisms are usually microscopic in size but there are notable exceptions such as the macroscopic brown alga, *Sargassum*. This very large plant lives in the open ocean. Its natural home is along the shore, attached to the bottom, whence each individual is torn.

The *phytoplankton*, or plant plankton, is the basic foundation of the pelagic *biota* (Greek, living things). This drifting flora synthesizes its food by the process known as photosynthesis and transfers its energy content as it is consumed by the animals.

Photosynthesis is a chemical process which transforms carbon dioxide and water to carbohydrate, or plant substance, while releasing oxygen in accordance with the following chemical equation with chlorophyll as the necessary catalyst:

$$6CO_2 + 6H_2O \xrightarrow[\text{Sunlight}]{\text{Nutrients}} C_6H_{12}O_6 + 6O_2$$

Only 0.02 percent of the total received sunlight is converted by photosynthesis into stored energy on a clear day. The reverse process, *respiration*, is the plant function at night. Carbon dioxide is released at a rate of only 10 percent of that used during maximum photosynthesis.

With photosynthesis being the only major source of oxygen to the world, and the only natural method of energy conversion to food substance, one can readily understand why the plants are such valuable links in the energy chain of life.

Since plants are being constantly consumed and destroyed, plant production to provide a standing crop increase must be prolific above the *compensation depth*, that depth at which oxygen utilization and oxygen production are equal. This is true since there is no significant plant production below the compensation depth. The compensation depth depends on the season, time of day, cloudiness factor, and water character. Recall that oxygen can be provided from the atmosphere by mechanical turbulence and surface cooling; this adds to the oxygen supply in many areas.

Because only two percent of the sea floor is lighted enough to support attached plant life, the importance of phytoplanktonic development is paramount. The primary nutrients: *nitrogen*—usually in the form of nitrates, *phosphorus*—usually in phosphates, and *silicon*—in the primary silicate form, must be present for plant development in the lighted depths.

The nutrient cycle is complex but can be basically traced with the seasons. In the surface layers during the period of reduced winter sun, there is a dearth of planktonic life. The nutrients are being released by decomposition and are stored in readiness for the spring bloom. As the sun rises with spring, there is a phenomenal increase in planktonic activity; the plant nutrients are quickly exhausted in the surface layers. The water color changes from a deep clear blue (the oceanic desert color) to a brownish green as the planktonic numbers multiply. As June approaches, the surface water has been warmed to the extent that convection from below by normal wind turbulence is stopped, and the nutrients cannot be replaced. The bloom slows considerably. In the fall as the storm activity increases and the surface water becomes less stratified, nutrients are again brought to the surface. As a result, a secondary bloom occurs since there is sufficient sunlight to sustain production to a measurable degree. Then the winter biological inactivity sets in and the cycle is complete. There is sufficient CO_2 being dissolved from the atmosphere at all times.

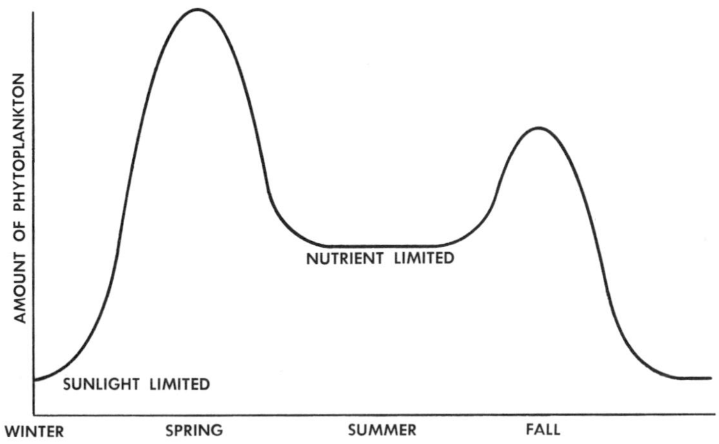

Figure 14-2 Seasonal Variation of Plankton Bloom (Photosynthesis)

The deep-ocean basins are storehouses of nutrients. Decomposition and oxidation of organic materials, as they filter down, rob the deep ocean of its oxygen which must be renewed to have life continue. The obvious extreme may be observed in the North Pacific Ocean. Its deep regions are extremely rich with nutrients while having a low oxygen content. This combination exists as a result of the lack of deep circulation.

The phytoplankton with the greatest number of species (20,000) is the *diatom* (Greek, cut in two). While generally a single-celled individual, it can exist in a series of linked cells to create strange and distinctive appearances. The shell or *frustule* is composed of translucent silica which is of considerable importance in siliceous sediments around Antarctica and under areas of significant upwelling. (Great diatom fossil deposits known as diatomaceous earth are used in naval boilers as primary firebox insulation.) Its reproduction is by simple cell division which occurs in enormous numbers during the spring bloom. It prefers cold water but is found in all localities.

The second most important phytoplankton in the ocean is the *dinoflagellate* (Greek, *dinos*, a whirling, and Latin, *flagella*, whips). Some species behave like animals in their food ingestion processes but as a group are better considered with the phytoplankton because of their photosynthesis faculties. In many areas they are the primary group com-

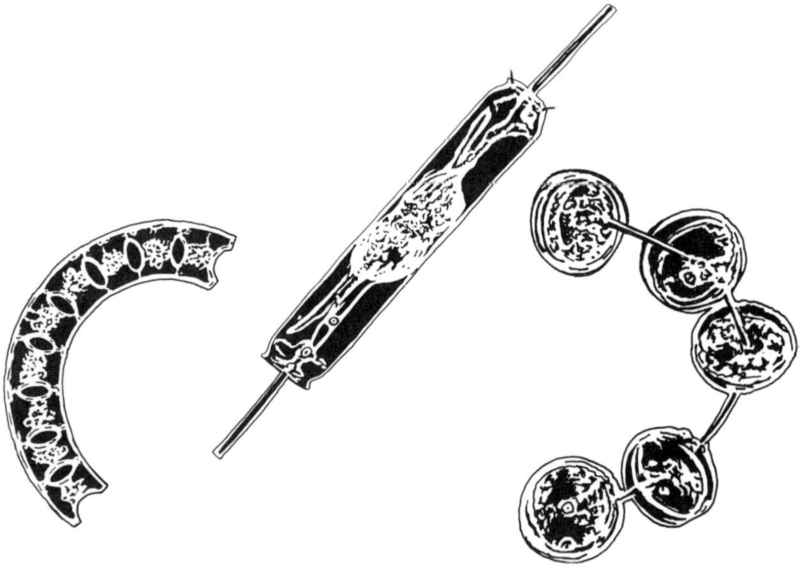

Figure 14-3 Characteristic Diatom Samples

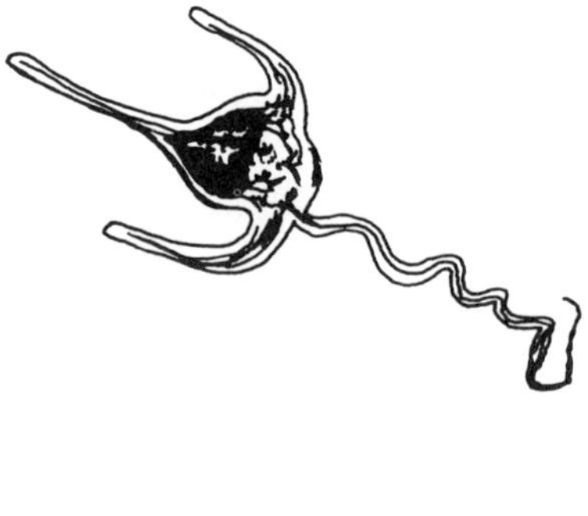

Figure 14-4 Typical Dinoflagellate

posing the disastrous "red tide" at "bloom" time which causes the coastal waters to appear as reddish brown soup. The "red tide" sometimes is thought to cause large fish kills by excreting a toxic gas to the water. The result is mass suffocation and temporary fisheries destruction. *Diatoms*, which are algae, and *coccolithophores*, which are flagellates, are also present in many of the "biologic tides."

1405 Zooplankton. Zooplankton refers to the animal part of the planktonic world and depends primarily on the phytoplankton for its sustenance. The majority are microplankton but some species are quite large, such as the Portuguese man-of-war.

The *copepods* (Greek: oar and foot), crustaceans, comprise 70 percent of all the zooplankton. They range in size from 0.3 mm to 8 mm and gather food with fine bristles on certain appendages. Copepods are otherwise known as "filter feeders," a common term which describes the forms that feed upon microscopic organisms and suspended *detritus* (organic debris). Small fish find them excellent food.

Figure 14-5 Copepod and Foraminifera

Foraminifera, protozoa, which include the abundant smaller *globigerina* are important for their calcium carbonate shells. After death, the foraminifera ooze on the bottom is useful in the study of past geological conditions and present petroleum geology.

Radiolarians, also protozoa, are minute planktonic animals with silica shells. The shells have produced the siliceous oozes which predominate the bottom of the Equatorial Pacific Ocean.

1406 Deep Scattering Layer (DSL). Some members of the zooplankton community have become light-affected or *phototrophic*. Many of the species have been observed to live in layered colonies which move toward the surface at night and dive to darkened depths during the day.

The first concerted study of these layers began when they were discovered by the U.S. Navy in connection with the underwater acoustical research performed during World War II. When highly concentrated, the layers present a false depth to the precision depth recorders and they have occasionally been mistaken for seamounts. They are now a potential shield to submarines at night and, as the operating depths of submarines increase, the layers will increasingly affect detection during the day.

Siphonophore have been identified as major members of the DSL society. These free-swimming or floating pelagic forms are usually delicate, transparent, and colored. Others listed as present are the shrimp, copepod, squid, and such deep-sea fish as hatchetfish attracted by the plankton. The DSL often appears as three distinct layers at midday, but becomes diffuse during the night. The daytime depth is between 700 feet and 2,400 feet—the deepest always at local apparent noon on a clear summer day. The shallowest depths occur on a mooonless night just below the wind-affected surface.

1407 Bioacoustics. Just as many marine animals reflect or scatter sound, there are many that generate it. Sounds produced by animals fall into two basic categories: those produced by an organ within the animal's body, and those produced by rubbing, cracking, or gnashing of shell-like extremities. The latter are exemplified by sounds produced by the crustaceans such as shrimp, lobsters, and crabs which either snap their *chela* (Greek, claw), as in the case of small snapping shrimp, or rub their legs together. Squid and octopus, on the other hand, produce a metallic rasping noise when they feed. Sometimes these noises become quite great, especially in the case of snapping shrimp, which are fairly common in coastal areas of warm seas throughout the world. From Table 8-3 one may see that this sound is equivalent in acoustic pressure to a loud home hi-fi set.

Fish, too, can make noises of this type. Some fish produce sounds when they gnash their teeth, perhaps during feeding, and other times perhaps when frightened or in serious difficulty. Fish also may produce sounds by oscillating the swim bladder (consisting of a sac of gas) which is basically used for buoyancy purposes. Apparently though, the swim bladder may also be used as a resonance chamber to produce all sorts of noises, especially of a foghorn type. Croakers, sea trout, sea drum, sea bass, marine catfish, toadfish, and squirrel fish are some examples of fishes which produce calls of this nature.

Of course, marine mammals such as the dolphins, porpoises, and whales also produce sounds. These sounds generally fall into two groups: one a clicking, whose rate may be varied by the animal; and the other, a whistle or squeal. The clicking sound apparently is used to locate objects by returning echoes, in other words, sonar. As a matter of fact, the bottle-nosed dolphin is very much adapted for this, having an oil-saturated, melon-shaped organ in his head for focusing the emitted acoustic signal. The returning signal is received by two ears acoustically matched to the water which achieve their directional properties from being located on each side of the head. Dolphins also emit high-pitched whistles and squeals varying from 1,000 to almost 20,000 hertz, which are evidently utilized for some sort of communication between members of the same species. Even the large whales produce sounds, some of them very low in frequency.

Whales being very large animals, even the heartbeat of the larger ones is apparently audible. Sounds having a frequency around 20 Hz and an intensity 30 to 40 db above ambient levels have been associated with this source. This seems plausible, because the heart of a blue whale weighs about 1,000 pounds and evidently pumps enough blood to expend energy at a rate close to 10 horsepower.

1408 Benthos. The benthic world includes all life on or in the bottom of the seas. The organisms are either *creeping*, *sessile*, or *burrowing* in their modes of existence and are present in varied forms from the intertidal zone to the hadal-benthic depths.

The *creeping* forms such as lobsters, crabs, snails, and some fishes *walk* or bounce on the bottom. All creepers are animals.

The *sessile* (Latin: *sedere*—to sit) benthos are firmly attached to the bottom. The sessile plants such as all the seaweeds and eel grasses and many of the diatoms must all be in the euphotic range for their very existence. Those in the intertidal zone vary greatly in characteristics depending upon the tidal range (commonly 15 feet in many areas). The sessile animals such as the sponges, barnacles, oysters, and corals are generally filter feeders. As such, they exist most commonly in the euphotic zones close to the greatest food supply. The coral specifically has been directly responsible for building islands simply because of its affinity for light. In many cases, skeletal coral remains have been deposited faster than mid-ocean seamounts have sunk, and thus an atoll was created. Other sessile animals are detritus or carnivorous feeders and can live in greater depths.

The *burrowing* organisms hide from their predators by digging into the mud, rock, or wood. They are represented by the clams, worms, and some crustacea. The most notable for the marine interests are *teredos*, ship worms that gnaw wood, and *limnoria*, isopod crustaceans, both of which bore holes.

Table 14-1 Burrowing Organisms

	TEREDOS (Teredo Novalis)	LIMNORIA (Limnoria Lignorium)
1. Method of Attack	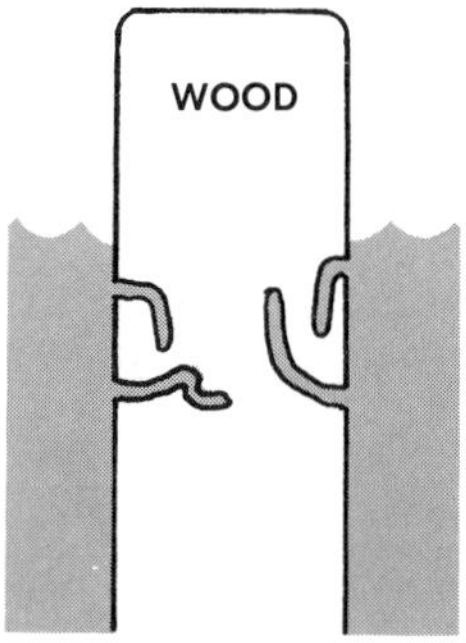	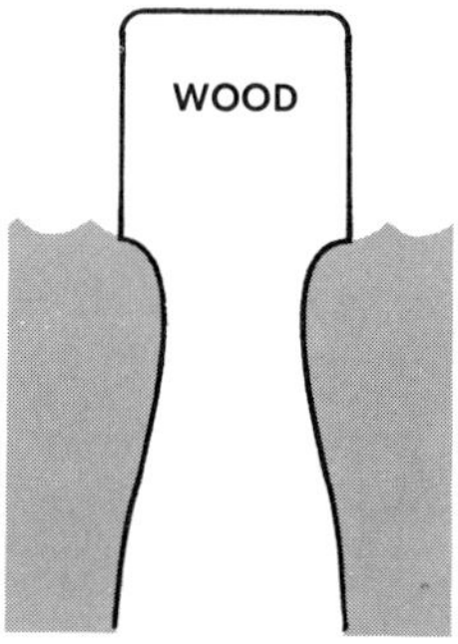
2. Environmental Effects	General Comment: These effects may be noticeable only after 3 weeks of exposure.	
a. Salinity	Activity drops at 9‰ LETHAL Below 5‰	Activity drops at 12–16‰ LETHAL Below 9‰
b. Temperature	Both can attack in adult form at 1°C; they then become dormant until temperature increases.	
c. Food	Plankton; cellulose required for metabolism process.	Cellulose (Wood)
d. Pollution	Both seem very immune to normal harbor pollution.	
e. Currents	Attack possibly deterred by speeds greater than 2 knots.	

1409 Boring and Fouling Organisms. The benthic organisms of most concern to the marine community are those that burrow into hulls and piers and those that encrust all marine surfaces. Both the boring and the fouling organisms are a constant problem; each category is discussed below.

The two most prevalent borers are teredos and limnoria, the prime destroyers of wooden piers below the high-water mark. Their methods of attack and their reactions to the environment are compared in Table 14-1.

There are several measures used to protect the wooden surfaces of piers from the borers. Pilings properly creosoted can last for many years.

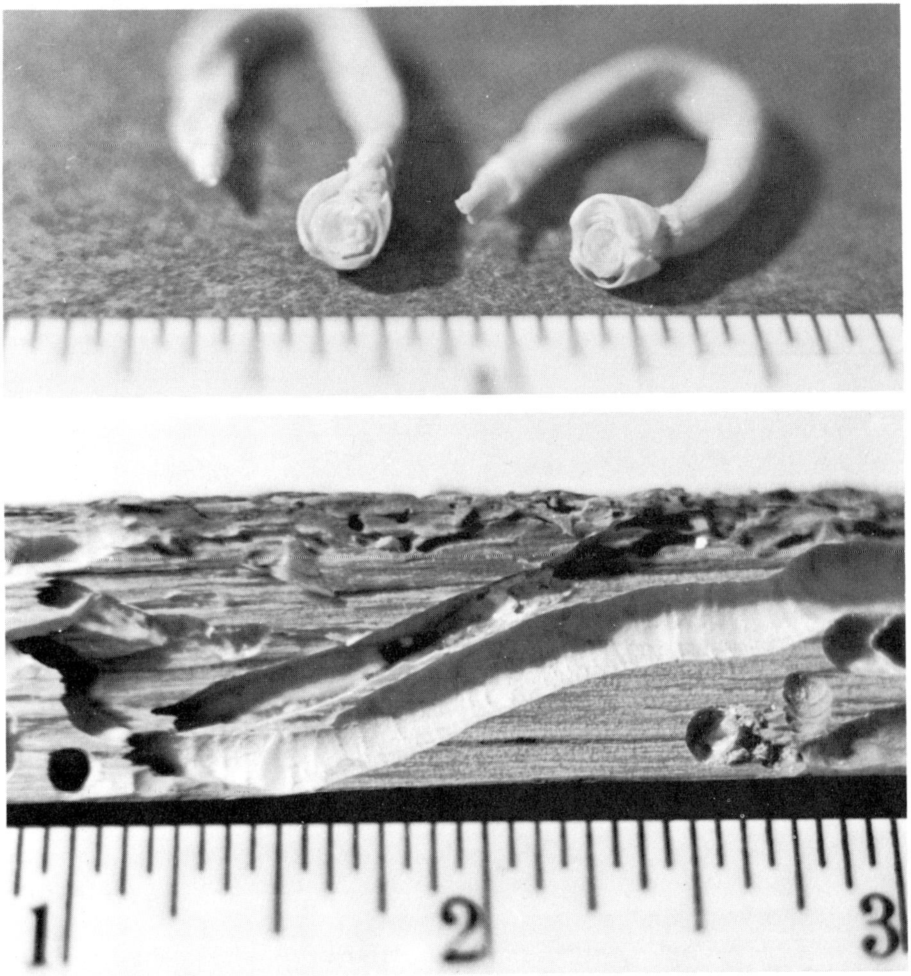

Figure 14-6 Teredo Specimens and Cross Section of Damaged Plank

Figure 14-7 Primary Limnoria Destruction

Placing concrete jackets over the core of wood used for pier construction provides more permanent protection, but the process is much more expensive than creosoting plain pilings. The most expensive method is to use reinforced concrete below the high-water line.

Some wooden hulls are covered with an outer wooden sheathing, which is considered expendable, but if it is properly painted and periodically cleaned, it will never have to be replaced. Extreme care should be taken to prevent the teredo from attacking the inner hull.

One problem is that of detecting the boring organism since, in most cases, fouling organisms such as barnacles will tend to cover up any evidence of deterioration. Sonic methods have recently been developed, however, which can readily distinguish between solid and hollow wood.

The fouling organisms are many and varied. They constitute the marine growth, plant and animal, attached to submerged objects. They cling to a new surface immediately; diatoms and other algae may have a 2,000 per square centimeter density within 48 hours after submersion. Protozoan creatures feed and then cling after about one week. Navigational buoys have been observed to collect 25 pounds per square foot of fouling organic material.

The effects of fouling are direct and serious. Propulsion efficiency for ships is seriously curtailed within a year of docking.* Oceanographic

* Process of removing ship from water for inspection and repairs.

instruments become inoperative. Fouling organisms will destroy anti-corrosive coatings on hulls and will seriously affect the sonic properties of sonar transducers and echo sounders.

Antifouling paint is now the most efficient covering to prevent growth, although it is unfortunately not completely effective. Cleaning (including blasting or scraping) and painting restores the smooth surface required for efficient propulsion.

1410 Nekton. The *nekton*, or free-swimmer, has the ability to move with controlled locomotion over considerable distance. This ability has apparently developed to allow escape from enemies, to pursue its prey, and to negotiate necessary migratory travel. Study and observation have conclusively shown that the nekton is unalterably dependent on other life for its existence; it is assumed, therefore, that plankton and benthos were fairly well established in the evolutionary chain before the nekton developed ecologically.

Nekton is primarily composed of fish, both demersal and pelagic; invertebrates such as octopus and squid; and marine mammals.

The *demersal fish* are those which spend their adult life swimming very close to the bottom, usually in the neritic province, and have adapted their physical characteristics accordingly. Such changes as having two eyes on the upper side of the head and coloring to blend with the bottom, are notable in the category of flat fish. There are also round demersal fish which are less developed to the particular needs of bottom living. Some primary examples of the demersals are flounder, cod, haddock, sole, and halibut, all of which have commercial importance.

The pelagic fish generally live near the surface in both the neritic and oceanic provinces wherever there is a sufficient planktonic food supply. The sharks and true *fish* (water-breathing through mouth and gills and having bony skeleton) are both travelers of the open sea. Each species has adapted itself to the primary environment of preference, but does, in many cases, migrate great distances for spawning. The eel and salmon are examples. Much of the profitable fishing is in the cold-water regions but there is also a tropical pelagic fish community which is now being exploited.

The squid has a water-jet propulsion mechanism making it unique in the marine environment. It operates in a large depth range and is considered a table delicacy in countries other than the United States.

The marine mammals are thought to have moved from the dry land since their original development. The whale is joined by the porpoise and dolphin in the nektonic realm. The seal, otter, and polar bear are also properly considered as nekton, but must live near the shore to care for their young.

There has been much research conducted with the playful porpoise, most of it sponsored by the U.S. Navy. Porpoises are easily trained and are known to possess unique acoustical abilities to communicate. In a recent Sea Lab session, a tame porpoise reliably delivered the mail from the surface to the oceanauts. A surface similar in nature to the porpoise's *epidermal* texture has been applied to submarine hulls in an effort to copy the minimum skin friction experienced by the animal.

1411 Food Chains. A *food chain* is the transfer of food energy by organism consuming organism. There are three different food chains: the *predator* chain, the *parasite* chain, and the *saprophytic* chain.

The first chain type, the *predator chain*, proceeds as expected from the phytoplankton, the *herbivore* (plant eating), and then to the various

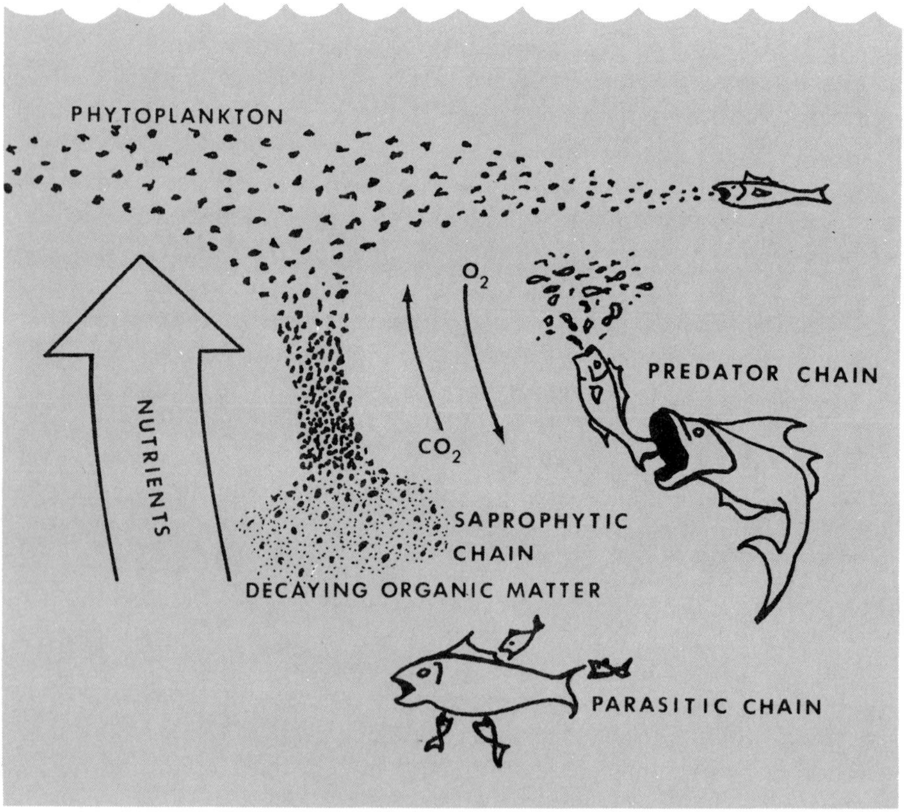

Figure 14-8 Food Chain and Nutrient Cycle

increasing sizes of *carnivore* (flesh eating). Each level is known as a *trophic* level or level of similar size. The predator chain most obviously is the one in which man participates in his search for food.

An interesting departure from the gradual size increase is the feeding method of the right (baleen) whale. It filters vast quantities of water through the *baleen plates* in its mouth to take *krill*, a small (1″–2″) zooplankton, directly for its sustenance. This is extreme efficiency, bypassing other energy-using trophic levels. A familiar analogy of such efficiency is the cattle's grazing on grass.

The *parasitic chain*, the second chain type, begins with a large individual with each succeeding trophic level being decreased in size.

The *saprophytic* (Greek: rotting plant) *chain*, the last type, goes only from dead organic matter to micro-organisms such as bacteria. The saprophyte, a heterotrophic vegetable organism, lives upon dead organic matter. The three food chains have developed in perfect harmony to maintain the necessary environmental balance. It behooves man not to interfere with this equilibrium.

1412 Bioluminescence. Bioluminescence is the emission of light by living organisms. The process is thought to be extremely efficient; since the light is radiated, very little heat, if any, is emitted. This is in great contrast to the electric light, heat energy releasing process. The phenomenon is chiefly an animal production and is widely distributed throughout the kingdom. At present, approximately 240 genera (groups) of organisms have been identified as bioluminescent but many more are expected to be observed. Common organisms of this type are dinoflagellates, jellyfishes, copepods, euphausiids (small shrimp), squids, and some fishes.

Bioluminescence results as the substance, luciferin (Latin: light bearing), is oxidized in the presence of the catalytic enzyme, luciferase. The exact chemical relation is very complicated and is presently under investigation. Bioluminescence is triggered by unexplained internal actions or external actions such as surface and internal waves; ship, fish, and whale movements; and upwelling. At night, ship wakes and regions of plankton bloom are easily recognized. The phenomenon occurs more in warm tropical waters, but is also found in northern waters during the warm seasons. It also seems to be more prevalent in coastal regions near river mouths and nearshore upwellings than in the open sea.

Bioluminescence is nature's demonstration of a potential method of energy production which may be industrially developed in the future. The efficiency of the process is excellent and possible applications such as safe deep submergence light are worth the time and money now being expended on basic research.

1413 Dangerous Marine Animals. No one questions the fact that a shark can disembowel a swimmer with one snap of the jaw. Many have witnessed or experienced jellyfish stings while swimming. Some have read or heard about the terrible cuts which may be sustained by scraping a coral colony. But few persons know what aftereffects would follow the eating of a tropical fish in time of need.

Representative examples of each major dangerous category are presented below. The categories are those that *bite*, *sting*, and are *poisonous to eat*.

Figure 14-9 Killer Whale in McMurdo Sound

The killer whale, representative of the *biting animals*, will attack anything that swims. It is present in all areas from tropical to polar regions. It travels in 3- to 40-member packs and lives by preying on other warm-blooded marine animals. The killer whale is identified by a white spot above and aft of its eyes. Its body is generally jet-black with very white underparts. When the whale is detected, the swimmer or diver should retreat immediately.

Stinging animals are generally less dangerous than biting animals but are more numerous. *Coelenterates* (Greek: *koilos*—hollow and *enteron*: an intestine), which include the common jellyfish, sea anemones, and corals, abound in the oceans. Various species of jellyfish are everywhere; their hanging nettles sting but rarely have more serious consequences.

Benthic anemones, such as the actinia, can also give painful stings when touched. Corals in the tropical waters are stingers in that their shells will produce ugly cuts. The stinging aspect is the poisoning which usually accompanies the wounding and inhibits the healing process.

Some animals which bite are classed as stingers if they inject toxic venom during the interval of physical contact. Examples are the moray eel and the sea snake, the latter being indigenous only to tropical seas. The snake's venom is considered to be the most toxic of all reptiles'; there are no known specific antiserums.

Non-edible organisms are divided into two major groups—*molluscs* and *fishes*. The molluscs (Latin, soft), popularly called shellfish, are generally safe to eat but may be rendered periodically dangerous when contaminated by Red Tide. Consumption of the dinoflagellates comprising the Red Tide makes the mollusc meat temporarily toxic.

Poisonous fishes have been difficult to identify because the toxic aspects of the fish meat has an unexplained periodicity. Toadfish and pufferfish meat except for the liver and intestines are generally safe, but even so, severe sickness has resulted from eating meticulously selected pieces. Also parts of some mussels such as the gill, liver, and intestines should always be avoided.

Shark and moray eel meat should be avoided if possible; although some species of both are safe. Treatments are not specific for meat poisoning; however, try to empty the stomach as soon as possible if poisoning is suspected.*

1414 Instruments. The primary criterion of biologic instruments is the retrieval of organisms in their original unharmed state. *Nannoplankton* (Greek: *nanno*—dwarf), must be centrifuged because it is too small to be caught by the smallest mesh. Various mesh nets with preset depth-keepers catch larger plankton and tiny fish. Ordinary nets are used for the larger pelagic specimens.

Benthic fauna are retrieved with trawls and dredges presently in a random fashion. Much organic destruction results from the heavy, clumsy equipment; the man-in-the-sea programs will be of the greatest assistance in this field.

Corers are the best devices available for collecting undisturbed samples. Tiny living organisms in the mud can be brought to the surface embedded intact in their natural muds.

1415 Commercial Fisheries. The fishery industry in the United States in its entirety added about $2 billion per year in recent years to

* For greater detail and excellent descriptions see *Dangerous Marine Animals*, by Dr. B. W. Halstead.

Figure 14-10 Isaacs-Kidd Midwater Trawl

Figure 14-11 Shallow-Water Plankton Net

the national economy. However, the total value of the catch for the fishery industry was only about $450 million in 1967, excluding the value added for processing, packing, and distributing. The annual per capita fish consumption of eleven pounds supported this commerce; but after considering that Europeans on the average consume over thirty pounds per person and that 80 percent of the total protein diet in parts of Asia and the Arctic is marine in origin, one can realize the true importance of the ocean's bounty now and its potential in the future.

Fish have been caught and eaten since pre-historic times. The abundance of fish was quickly recognized as one of North America's primary resources soon after Columbus' cruise. The nineteenth century with its scientific awakening brought with it the beginning of commissions and international conferences to study and protect the marine organic resources. Today there are many treaties such as the International North Pacific Fisheries Convention (member nations: Japan, Canada, and United States) to protect the salmon and the International Halibut Commission with Canada and the United States as members. Recently the Malta Resolution was proposed in the United Nations, which stated that all the fishing resources beyond the 12-mile limit would be under the ownership of the United Nations. Unilaterally, many countries have set a 12-mile limit for fishing and a three-mile limit for commerce.

There have been official discussions concerning the continental shelf ownership for legal exploitation of the bottom and geologic resources. The question of whether the lobster falls into this discussion has been argued and is presently unresolved. Peru is unilaterally claiming a 200-mile fishing limit and has by force of arms arrested several fishing trawlers outside the 12-mile line, but has remained publicly unchallenged at this writing.

Peru has led the world in the catch of fish for the last few years after a great leap in trawler and canning-facility construction. She is concentrating in the production of sardines, herring, anchovies, and tuna. Other leaders are Communist China, the Soviet Union, Norway, the United States, and Canada.

In recent years, Japan has consistently been among the fish-catch leaders in the world; this has been the direct result of a national awareness of the ocean's potential and the need to maintain a modern efficient fishing fleet to complement the total food production required to feed her people well. Japanese biologists are leaders in fields of shellfish farming and shallow-water fencing for transplant of fish species.

The present United States effort is dwindling. Many factors have caused a stagnation in this nation's efforts to commercially develop its fisheries. The ships are old and worn. New, efficient fish-factory trawlers can cost close to $10 million. Stern ramp, medium size (200′) fishing trawlers cost about $4 million. The Bureau of Commercial Fisheries is conducting, on a small scale, excellent basic and applied research in its several laboratories around the country. The academic community at present is placing very great emphasis in the fields of biological oceanography and marine biology. But the total effort is a very small fraction of the Gross National Product, an unfortunate circumstance when compared to the potential.

What is the potential? While not unlimited, the ocean's potential is considerably greater than present yield in view of the fact that present fishing grounds are primarily on or near the continental shelf regions. Areas such as the North Sea, the Grand Banks, and the wide Southeast Asia Shelf are heavily fished at this time, but the deep-water regions of divergence, current confluence, and upwelling are barely tapped. Fishermen now realize that nutrients at the surface bring the large fish schools. Heavy weather turbulence can draw fish to areas otherwise barren. Other nations are building fleets which can be self-supporting for long periods to hunt these areas and exploit them. The United States should attempt to do the same.

In early 1967 the Food and Drug Administration approved fish protein concentrate (FPC) for human consumption in this country. The FPC is a refined meal or flour which is made from the entire fish including parts previously discarded. Food economists insist the FPC is completely nutritious, and it is extremely inexpensive to produce in large quantities. The United States is attempting to project the use of FPC to help solve the world's food problems in Asia and Africa. An effective fishery effort to produce the catch for FPC production could increase the protein diet of the underfed to a level heretofore unforeseen.

Electric fishing methods used to concentrate schools of fish are being used to a small extent. The menhaden fisheries in the United States are using this method with success. Experiments with air bubbles in shallow water which form impenetrable barriers have assisted the herring industry off the coast of Maine. Methods of vacuum pumping have been developed to suck live fish directly into the trawlers' holds. There is also continued refinement of the common trawling nets and sound-ranging with sonar to locate the large schools. New equipment can accurately locate fish at two-mile ranges.

At present, fishermen in Southern California are transmitting weather data and BT temperatures to the Bureau of Commercial Fisheries in La Jolla in an experimental program. The data are transmitted by land line to Fleet Numerical Weather Facility (FNWF) in Monterey, California, which uses them in its oceanographic forecast. The Bureau of

Commercial Fisheries forecasts fish locations from the FNWF readout and then directs the fisherman to the most lucrative areas. The Navy is organizing a much wider program with all of the U. S.-flag fishing craft to obtain cheaply the badly needed data for its environmental predictions since *the data required to conduct fishing and antisubmarine operations are essentially the same.* The U. S. Navy and the Bureau of Commercial Fisheries are working together on co-operating cruises to identify various fish sounds by simultaneous use of sonar and fishing nets.

Estuary pollution is a problem now, and deep-water pollution is becoming a problem. The USSR has protested the pollution of the North Sea by the nuclear reactor plants in the United Kingdom. In the United States many commissions are studying how to combat the pollution which is killing the fowl wildlife as well as the benthos and nekton. Certain tributaries of the Chesapeake Bay are in immediate danger; many united efforts are now underway to improve the situation.

Commercial fishing may be the answer to the world's food problem. It is clear that the Federal Government must increasingly join the effort in assisting industry before it is too late.

1416 Man in the Sea. The aqualung development by the French diver Jacques Y. Cousteau in the 1930's has led to the amazing life-in-the-sea technology being developed today. Sea Lab II, lasting thirty days, accomplished all plans to investigate the human effects of the high pressures, special gas atmosphere, and unfamiliar habitat. Cousteau's *Conshelf* program in the Red Sea was equally successful. The submerged quarters were comfortable and the daily underwater routine was quite productive. Using exotic breathing mixtures, such as helium-oxygen combinations, modern experiments indicate that saturation diving can be performed over extended periods of time at depths greater than 1,000 feet. The entire continental shelf is now fair game for commercial ventures requiring human labor. Thus man is learning to cope with life on the ocean floor and is broadening his knowledge of the shelf regions.

Volunteers are already available to participate in the underwater direct-breathing program. The object is to discover how to simulate the gill function of the fish in order that man will be free of all oxygen gas sources. He would be able to explore unlimited areas should the program succeed.

Many immediate applications have evolved from the long-term efforts. Salvage is much easier now than it was thirty years ago, and complex repairs to oil wells can be made hundreds of feet below the surface. The biologic catalogue has greatly expanded because of the sportsman's adoption of scuba gear.

In short, man's present sea technology was considered fantasy 35 years ago; there now seems to be no limit to its potential.

Additional Reading

Barnes, H., *Oceanography and Marine Biology*, George Allen & Unwin Ltd., 1959.

Coker, R. E., *This Great and Wide Sea*, The University of North Carolina Press, 1947.

Dietrich, Gunter, *General Oceanography*, John Wiley and Sons, New York, 1963

Dietz, R. S., "The Sea's Deep Scattering Layers," *Scientific American*, August, 1962.

Fish, M. P. and Mowbray, W. H., *Sounds of Western North Atlantic Fishes*, The Johns Hopkins Press, 1970.

Gilbert, P. W., "The Behavior of Sharks," *Scientific American*, July, 1962.

Halstead, B. W., *Dangerous Marine Animals*, Cornell Maritime Press, 1959.

Herring, P. J. and Clarke, M. R. (Eds.), *Deep Oceans*, Praeger Publishers, 1971.

Holt, S. J., "The Food Resources of the Ocean," *Scientific American*, September, 1969.

Isaacs, John D., "The Nature of Oceanic Life," *Scientific American*, September, 1969.

King, Cuchlaine A. M., *An Introduction to Oceanography*, McGraw-Hill, Inc., 1963.

Long, E. J., Capt., USNR, Ret., Ed., *Ocean Sciences*, United States Naval Institute, 1964.

McElroy, W. D., and Seligner, H. H., "Biological Luminescence," *Scientific American*, December, 1962.

Pequegnot, W. E., "Whales, Plankton, and Man," *Scientific American*, January, 1958.

Russell, F. S., and Yonge, C. M., *The Seas*, 3rd ed., Frederick Warne Co., New York, 1963.

Slijper, E. J., *Whales*, Basic Books, Inc., New York, 1962.

Sverdrup, H. V., Johnson, M. W., and Fleming, R. H., *The Oceans*, Prentice-Hall, Inc., 1942.

Wald, George, "Life and Light," *Scientific American*, October, 1959.

Williams, Jerome, *Oceanography*, Little, Brown and Company, Inc., Boston, 1962.

CHAPTER FIFTEEN

Condensation and Precipitation

1501 Condensation. Most people have been annoyed by the salt that will not come out of the shaker during periods of high humidity or rainy weather. This is a manifestation of near saturation coupled with salt's affinity for water vapor. Normally, saturation is said to take place when the net exchange of water-vapor molecules across an air-water interface is zero, or when as many molecules go from the surface of the water into the air as leave the air and return to the water. This is the situation referred to as 100 percent relative humidity. Two factors affect this condition of equilibrium. One is the curvature of the air-water interface, and the other is the purity of the liquid water involved. These are known as the *curvature effect* and the *solute effect*. They will be considered separately.

The *solute effect* is such that the saturation vapor pressure over a solution can be less than that for pure water, depending on the strength of the solution—the stronger the solution, the lower the saturation vapor

Rain-spattered windshield of the bridge aboard aircraft carrier USS Ranger pressure. What will produce a solution? Something must be present in the atmosphere to serve as the substance to be dissolved. Salt fills the bill perfectly. Particularly over the oceans, spray puts large quantities of salt into the air and provides what are called *hygroscopic nuclei.* The term hygroscopic (Greek, *hygros*—wet + *skopos*—object) indicates that the particle has an affinity for water. Other substances over land such as the products of industrial combustion, particularly those of sulfurous or nitrous content, make similar suitable hygroscopic nuclei. Dust, smoke, and haze are ample evidence of the particles present in the atmosphere. The size of these nuclei ranges from 0.1 to about 1.0 micrometers in diameter. To put this in proper perspective, it should be noted that an ordinary cloud droplet ranges from 10 to 100 micrometers, while a fine rain may be 500 micrometers in drop diameter. A large raindrop may have a diameter of 5,000 micrometers. A nucleus, introduced into the atmosphere which is near 100 percent relative humidity, can take on water in liquid form. The discussion of saturation vapor pressure in Chapter 3 was limited to that of a gas, over a plane water surface. Once the hygroscopic nucleus has been introduced and a solution produced, a cloud droplet exists. Because of the water's unique quality of high surface tension, this drop forms a spherical shape. Now the consideration of vapor pressure must be examined over a surface which is no longer plane but curved. This *curvature effect* is such that it tends to oppose the aforementioned solute process. When the drop is very small, the surface tension causes it to behave much like the stretched skin of a bubble, opposing the entry of any more molecules from the exterior, and it is only the solute effect that allows droplet growth. As the droplet grows, the solution becomes weaker and weaker, until the drop comes into equilibrium with the surrounding atmosphere. A cooling of the atmosphere to *supersaturation*, or greater than 100 percent relative humidity can destroy this balance. (It has been shown under laboratory conditions that "pure" air, devoid of condensation nuclei, can be supersaturated to a point approaching 400 percent relative humidity.) Supersaturation in the atmosphere does not exceed 100 percent to any extent since sufficient condensation nuclei are normally present to prevent it. It is in this fashion that the drop can be caused to grow, increasing its radius of curvature, approaching a more plane configuration. This growth tends more easily toward the condition of equilibrium between water and air.

The mariner and the aviator must be particularly alert to the relative humidity. More often than not, the proximity of the atmosphere to saturation is expressed as the *dew point spread*, or the numerical difference between the ambient air temperature and the dew point temperature. Particular caution is necessary when this spread is small. This is the time when a small reduction in the ambient temperature can cause condensation, which results in haze, fog, or low clouds. This can make the approach to a destination or a transit in restricted water difficult and hazardous.

Condensation is manifested in four forms: *dew*, *frost*, *clouds*, and *fog*. The form that results is predicated on the circumstances present at the time of formation.

1502 Dew. Dew is formed when the temperature drops, normally at night. When the sun sets, the amount of radiation leaving the earth is much greater than that received. The resultant cooling allows condensation of water vapor upon contact with cool surfaces such as grass, bushes, and fences. This process is of vital importance to certain sections of the country. The nearshore areas of California depend almost wholly upon this nightly dew formation to provide the necessary moisture to sustain most plant life. The amount of annual rainfall is insufficient to support life, and dew supplies the lifegiving water to the vegetation.

For dew to form, three prerequisites must be fulfilled. First, the radiating surfaces must be sufficiently removed from the heat supply of the soil. Second, there must be high relative humidity in the layer of air adjacent to the condensing surfaces. Third, the layers of air above the surface layers must be low in water-vapor content, in order to allow sufficient terrestrial cooling.

1503 Frost. Frost is quite similar to dew in the formative process. The only exception or restriction is that the temperature of the object upon which the condensation takes place must be below freezing. Water vapor in the air goes directly from the vapor (or gaseous) state to the solid state. This phenomenon is called *sublimation*. The formation of frost should not be confused with the freezing of water in dew form. When this occurs it is referred to as *white dew*. Frost also is known by the name of *hoarfrost*.

1504 Clouds. A cloud is the result of the condensation of water vapor into small droplets. This takes place because a quantity of air has been lifted or caused to rise. There are several ways in which this can occur: *orographic lifting*, *frontal lifting*, and *thermal convection*. However, when the air reaches a temperature in its cooling ascent, such that its ability to hold moisture is sufficiently reduced, condensation takes place in the form of clouds. In 1803, an Englishman named Luke Howard produced a system by which clouds could be uniformly described. This system has been modified slightly over the years but has been adopted by the World Meteorological Organization (WMO) as the standard for observers. Prior to this time, the description of the wonders, the beauty, and the awesomeness of the cloud vagaries and forms were left to the poet and the artist. In 1957 WMO published the International Cloud Atlas, a two-volume compendium with pictures, characterizing the genera, species, and special categories. There are ten genera, fourteen species, nine varieties, and nine supplementary features.

The basic features of the system are shown in Figure 15-1. It can be readily seen that the three basic genera are *cumulus*, *stratus*, and *cirrus*. These clouds are further subdivided into the general altitude levels where they are found: high, middle, and low. High clouds are those found between 16,500 feet and 45,000 feet. Middle clouds occupy the altitudes between 6,500 and 23,500 feet. Low clouds are found below 6,500 feet.

Table 15-1 Cloud Classifications
(Middle Latitudes)

<table>
<tr><th>ALTITUDE</th><th>STRATUS (St)</th><th colspan="3">CUMULUS (Cu)</th><th>CIRRUS (Ci)</th></tr>
<tr><td>HIGH
16,500–45,000′</td><td>CIRROSTRATUS (Cs)</td><td>CIRROCUMULUS (Cc)</td><td rowspan="3">TOWERING CUMULUS</td><td rowspan="3">CUMULONIMBUS (Cb)</td><td>CIRRUS (Ci)</td></tr>
<tr><td>MIDDLE
6,500–23,500′</td><td>ALTOSTRATUS (As)</td><td>ALTOCUMULUS (Ac)</td><td></td></tr>
<tr><td>LOW
0–6,500′</td><td>STRATUS (St)</td><td>CUMULUS (Cu)</td><td></td></tr>
</table>

Notice that there is a certain amount of overlapping of the categories. Also, the figures are approximate and are given for middle latitudes. Since the shape of the atmospheric gas envelope is distended to relatively high altitudes over equatorial regions, a similar contraction is experienced over the frigid poles. A compensatory adjustment is made for categories of clouds that exist in these regions. Consequently high clouds are rarely found above 25,000 feet over the poles, whereas they may be found as high as 60,000 feet over the equator. Comparable adjustments are made in the other subdivisions of the polar and tropical regimes.

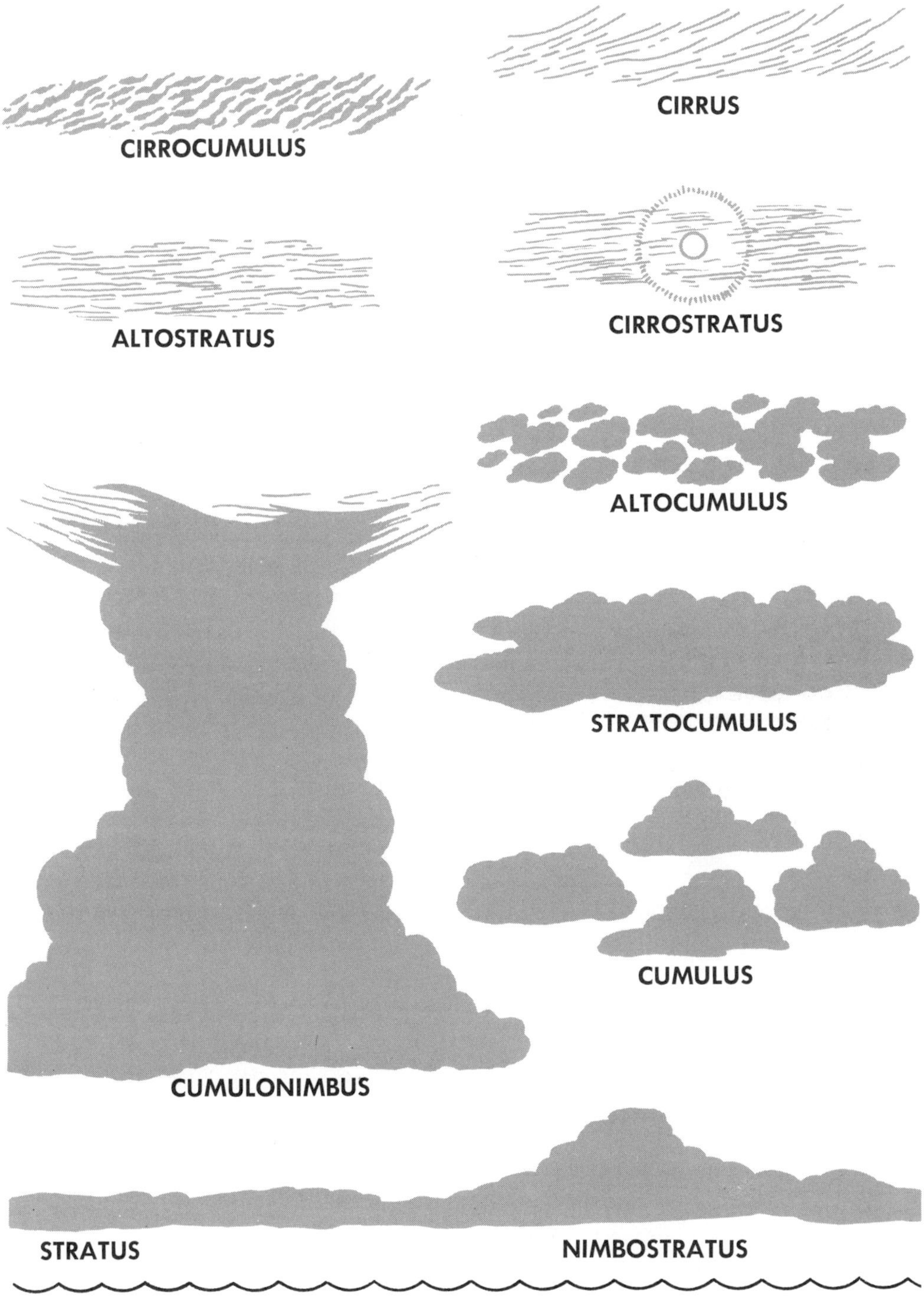

Figure 15-1 Typical Cloud Formations

Cirrus (Ci) are high clouds. Because of their extreme altitude they are composed almost exclusively of ice crystals. They have the appearance of delicate filaments or patches of white. Their structure is such that they appear to be composed of fibrous or hairlike strands. They are often called "mare's tails." Combinations of the other genera may produce cirrostratus or cirrocumulus. As one would expect, they have the characteristics of both type clouds, and appear at high altitudes.

Stratus (St) are low clouds whose appearance typifies their Latin root (the past participle of the verb *sternere* which means to flatten out or to cover with a layer). They cover large areas with a uniform fog-like texture. Their base is usually horizontal and uniform but not touching the ground.

The prefix *alto* denotes a middle cloud. The two basic genera, cumulus and stratus, at middle layers, are called altocumulus or altostratus.

Cumulus (Cu) clouds are dense, non-fibrous, vertically developed clouds with rounded dome-like tops, that appear much like the shape of cauliflower. Sunlight makes these clouds brilliantly white. These puffy clouds are characteristically present during fair weather and caused by the great vertical motion of midday heating. Frequently cumulus clouds will show considerable vertical extent, spanning the altitude range of both low and middle clouds. These are referred to as *towering cumulus.*

The *cumulonimbus* (Cb) are the truly spectacular clouds of the tropical and temperate belts. They are the nemesis of all pilots. The vertical motions which produce these dark towering giants can throw an aircraft out of control and cause severe structural damage. They are almost always accompanied by showery precipitation, lightning, thunder, and often hail and snow. These mountainous clouds can tower up from two to five miles high. At the mature stage, their tops are flattened and drawn out to an anvil shape. These are the *thunderheads*, which can darken the sky for miles around and from whose ragged base come enormous quantities of precipitation.

Other species of clouds bear the descriptive prefixes or suffixes such as: *fractus* (broken), *congestus* (piled up), *floccus* (fluff or tuft), etc., in combination with the basic generic names. These serve to better delineate the many and varied forms which these water droplets can assume.

Clouds in combination with wind information can give good short-term forecasts. With a little practice the seafaring man can become quite adept at using his knowledge of cloud forms and their normal sequences to make intelligent operational decisions. It is little wonder that the sailor of the past had become expert in mere survival because of his attention to the behavior of the environment. In these days of sophisticated and remarkable electronic and scientific devices, the aspiring mariner is often insufficiently aware of the environmental changes that surround him. A basic knowledge of the behavior of clouds would stand him in good stead.

1505 Fog. A discussion of clouds would be incomplete without consideration of that cloud whose base rests upon the ground—*fog*. Fog is always caused by the ambient temperature and the dew point temperature that are coincident or nearly so. Fog is rarely formed when the reported dew point spread is 4°F or greater. Fog can be formed by one of two basic processes: *cooling* the air to its dew point, or *increasing the moisture content* to saturation. In either case, condensation nuclei must be present in sufficient quantities. By definition, fog decreases horizontal visibility to 1 km (5/8 mile) or less. In order of increasing visibility ranges one encounters *fog*, *mist*, and then *haze*.

Fog formed by cooling air to the dew point is classified into three common types predicated on how this cooling is achieved. These are: *radiation*, *advection*, and *upslope*. *Radiation fog* is principally a nighttime effect. The cause is the cessation of incoming radiation from the sun. This allows a cooling of the earth and the adjacent layer of air. If the cooling is sufficient to lower the temperature to the dew point, fog will form. A gentle breeze which aids in mixing the saturated air will intensify the fog. Since the radiative cooling of the earth is the causal factor, clear, relatively dry air is necessary, in the layers adjacent to the ground layer. This removes the greenhouse effect allowing the heat to escape to the upper atmosphere. In addition, the presence of radiation fog and its

Courtesy: Approach

Figure 15-2 Final Approach to a Fog-Enshrouded Runway

attendant dry clear air at upper layers is usually an indication of good weather for the next day.

Many people say that the fog will dissipate by "burning off." They give the impression that, with the appearance of the sun the next morning, the fog is dispelled from the top down. This is not the case. One need only consider the process under which it formed to realize that the reheating of the earth, and its adjacent layer of air will cause the fog to "lift." This is the origin of the term *high fog* which is so prevalent on the West Coast of the United States. Since the fog is such a persistent phenomenon, particularly during the winter months, with a temperature inversion to hold down the condensed moisture, it alternates diurnally from fog to a low stratus layer (high fog) for days at a time. Since this fog forms next to the ground it is often called *ground fog* although technically ground fog, by definition, indicates that it obscures less than 0.6 of the sky and does not merge with the base of any cloud layer that may exist above it.

Advection fog depends upon the principle of bringing warm moist air over a cold body, usually water. Note the term advection implies horizontal transport of a property by mass motion (although strictly speaking, advection can have a vertical component, normally vertical motions are thought of as convection.) This is a common situation at sea, or along coastlines where upwelling brings large areas of cool water up from below. The moisture-laden air, over cold water, is quickly cooled to dew point. The Grand Banks of Newfoundland offer a prime example of this process. The warm moist tropical air, under the influence of high-pressure-induced motion, travels northward and encounters the cold Labrador current. Extremely thick fogs are formed there, creating severe hazards to sea traffic between North America and Europe. The combination of this fog and the increased incidence of southward iceberg drift makes this a particularly difficult area to transit, especially in early summer. It is interesting to note that advection fog depends on wind for motion of the moist air, while only a gentle breeze will sustain radiation fog, a stronger wind will prevent it, or dissipate it once formed.

The term advection fog is often used in another situation, yet the processes at work are slightly different. True, it does require the advection of cold air over warm water, but in this case the cold advancing air reduces the temperature of the already saturated, or nearly saturated, air adjacent to the surface, by mixing. Now the moisture that has just evaporated from the surface, is recondensed in a small layer near the surface. Because this process often occurs in high latitudes when very cold air is advected from over cold land or ice, it is called *Arctic sea smoke*, or *sea smoke*. It is a true *steam fog* which is normally not over 50–100 feet thick, but can be a hazard to ships or landing aircraft.

Another way of cooling air to the dew point is through adiabatic expansion. This can easily be done by lifting the air up a gentle sloping area of terrain. This is the formation process for *upslope fog*. A classic example of this situation is the fog formed by the gentle lifting of heavily moisture-laden air from the Gulf of Mexico as it travels northward across the Great Plains of the United States. This can cause lifting from sea level a 5,000-foot elevation. Here is a case of fog maintained by relatively strong winds, although very strong winds will cause turbulence and the formation of low stratus.

The second fundamental process for fog production, as stated before, is to increase the water-vapor content of the air until it reaches saturation. This can be accomplished by rain falling into a stratum, or wedge, of cold stable, nearly saturated, air. This necessary position of two air masses can occur at the front or boundary between them. This point will be mentioned again in the discussion of frontal weather.

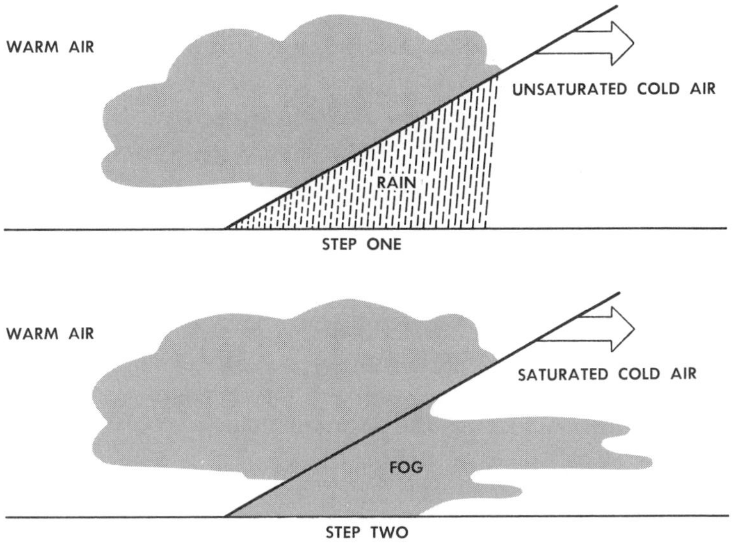

Figure 15-3 The Formation of Frontal Fog

1506 Precipitation in General. The process of condensation in which water vapor is formed into cloud droplets is quite familiar. The next step in the hydrologic cycle is not so well understood. *Precipitation* is defined as any or all forms of water particles, whether liquid or solid, that fall from the atmosphere and reach the ground. All forms of precipitation are included within the general classification of *hydrometeors*. There are about fifty different types of hydrometeors, but only the most familiar will be discussed here. *All precipitation falls from clouds, but not all clouds release precipitation*. It is believed that the cloud droplets in a cumulus last but a short time, on the order of minutes, before they are evaporated, while the droplets in a stratus deck may persist for an hour or two. The process of condensation must therefore be continuing, if the cloud is to persist for any extended period.

The growth of a cloud droplet to precipitation size is, on the other hand, dependent on many factors. Two major theories as to the growth of droplets to sufficient size are popular. The first is the *ice crystal* method, which was promulgated separately by two men, Tor Bergeron of Sweden and Walter Findeisen of Germany, and is at times referred to by this combination of names. The second has been designated the *capture* method, which is descriptive of the process.

1507 Ice Crystal Method. The *ice crystal method* is predicated on the fact that the vapor pressure over ice is less than that over water. Normally, with the adiabatic cooling that occurs with an increase of altitude, condensation is reached. An increase in height normally results in the supercooling of water droplets. Between the range of 10° and −20° Fahrenheit the composition of a cloud is usually a combination of ice crystals and water in droplet form. This results in a saturation vapor pressure that is a combination of the two existing components. The air is not quite saturated with respect to a water surface, but it is slightly supersaturated with respect to an ice surface. The process that follows is that the water droplets evaporate, while water vapor is condensing on the ice nuclei. The net results are the formation of a drop heavy enough to overcome the updraft and the onset of precipitation.

This *Bergeron-Findeisen method* offers a suitable explanation for rain, where the temperature in the upper reaches of the cloud permits it to function. However, it was clear that this explanation would not serve in tropical areas. There is an obvious superabundance of rainfall, yet the temperatures at higher altitudes would not support the theory.

1508 Capture Method. The capture method is involved with the variable size of condensation nuclei and the inevitable production of a quantity of nuclei much larger in size than the average. By a process of collision, during both up and down motion, these droplets coalesce with other smaller drops. As their size increases, they begin to fall and by *direct capture* gather up all drops in their path of fall. In addition, as they gain speed, they leave an area of reduced air resistance behind them. This could be thought of as a slight vacuum, which attracts adjacent water droplets into the wake and the falling drop itself. This is appropriately called the *wake capture* method. This process continues until the drop reaches terminal velocity and a size such that it either

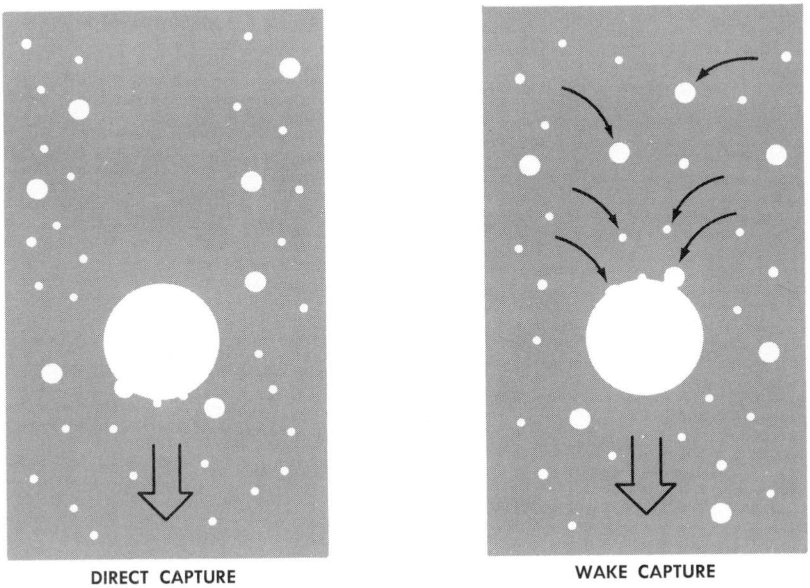

Figure 15-4 Capture Methods

reaches the ground, or is broken apart by the friction of the air. Thus the rain is formed.

Various combinations of temperature may modify the fall of rain before it reaches the ground. The size of the drop can be greatly reduced by falling through a layer of air that is of low humidity. The liquid water that does not reach the ground is called *virgae*, in fact, not meeting the criteria for precipitation, yet of such a similar nature as the items previously discussed, to suggest mention at this point.

1509 Precipitation Forms. *Snow* is the sublimation of water vapor which takes place at subfreezing temperatures. Snow crystals are hexagonal in shape and have been the constant fascination of all who see them magnified, because of their varied and beautiful structure. It is quite often the case that a cloud of large vertical extent will consist of water droplets in the lower levels and snow flakes at high altitude.

Sleet occurs when raindrops freeze or when largely melted snowflakes begin to refreeze prior to hitting the surface. *Freezing rain* is the term applied to a similar condition, except that freezing has not completely taken place before hitting the ground. In this instance, the rain freezes upon impact on cold objects; the resulting smooth coating of ice is called *glaze*. *Rime ice*, on the other hand, is formed when a moving object passes through air containing supercooled suspended water droplets. It is normally formed on ships and aircraft and has the appearance of a thick, white frost.

Hail is the result of a hydrometeor being alternately lifted above and falling below the freezing level. With each successive trip above the freezing level, the water that was acquired below is frozen. This process of concretion results in a number of concentric layers of ice similar in structure to an onion. Hail is the product of violent convective activity in cumulonimbus clouds.

1510 Weather Control. The idea that man could control the weather has tantalized meteorologists for many years. It is clear that there is a definite causal relationship between *precipitation* and the *nuclei* for condensation. Why is it that clouds will drift past dusty and drought-ridden farmland in the summertime, without sharing their much needed water? It has often been said that he who controls the weather can control the world. The aspects of defense use are numberless.

Many experiments have been conducted employing the theory of providing hygroscopic nuclei to clouds. A great number of them have been successful, or apparently so, yet the question often remains—"Did what was done cause it to rain, or was it going to rain anyway?" Dry ice, silver iodide, and large water droplets themselves have been used successfully to "prime the pump." Large-scale seeding operations involving areas of as much as a thousand square miles have been carried out with some success.

The prospect that an aircraft might be able to clear its own path for landing, to dissipate a fog, would be accepted with great enthusiasm by pilots and passengers alike. However, the facilities of all government agencies must be brought to bear on the general problem of weather control. Should it be made feasible, the legal aspects of the problem then take on alarming proportions. Does the farmer in an adjacent plot have the right to deplete the air of its moisture without sharing it with other neighboring farmers? And how about those who live twenty miles downwind? Consider the international complications in an area like western Europe with the close proximity of many nations sharing the same air mass. At least now Nature dispenses her favors with a record of impartiality that can be historically verified.

Additional Reading

Battan, L. J., *Radar Observes the Weather*, Anchor Books, Doubleday & Company, Inc., 1962.

———, *Harvesting the Clouds*, Anchor Books, Doubleday & Company, Inc., 1969.

Blanchard, D. C., *From Raindrops to Volcanoes*, Anchor Books, Doubleday & Company, Inc., 1967.

Byers, Horace, R., *General Meteorology*, 3rd ed., McGraw-Hill, Inc., 1959.

Donn, William L., *Meteorology*, 3rd ed., McGraw-Hill, Inc., 1965.

Huschke, Ralph E., *Glossary of Meteorology*, American Meteorological Society, Boston, Mass., 1959.

Petterson, Svere, *Introduction to Meteorology*, McGraw-Hill, Inc., 1958.

———, *International Cloud Atlas*, World Meteorological Organization, Atar, S.A., Geneva, 1957.

CHAPTER SIXTEEN

Air Masses, Fronts, and Pressure Systems

1601 Air Mass Formation and Identification. Just as a water mass may be defined in terms of a large volume of water having similar characteristics, an air mass may be similarly specified. An *air mass* is defined as a large homogeneous body of air which has approximately the same horizontal temperature and humidity characteristics.

By virtue of the fact that there are large areas of the globe which may have the same temperature and humidity characteristics, air masses are formed. The prime formation areas are those where a body of air can spend a considerable period of time. With sufficient time, the air will then take on the temperature and humidity characteristics of the land or water it overlies, achieving equilibrium with the surface. As a general rule, areas of high-pressure or anticyclonic winds are favorable for the formation of air masses. Figure 16-1 shows these source regions. They can be categorized as *Arctic*, *Polar*, *Tropical*, and *Equatorial*.

Bergeron, among his many contributions to the field of meteorology,

Cumulus cloud boils upward

formulated the classification system now used. He employed capital letters to roughly identify the latitude of the source regions: **A**-Arctic, **P**-Polar, **T**-Tropical, and **E**-Equatorial. To give some idea of the humidity of the mass, the lower case letter **c** or **m** is prefixed indicating respectively *continental* (dry) or *maritime* (moist). These letters used in conjunction with one another are sufficient to identify the mass *in its source region.*

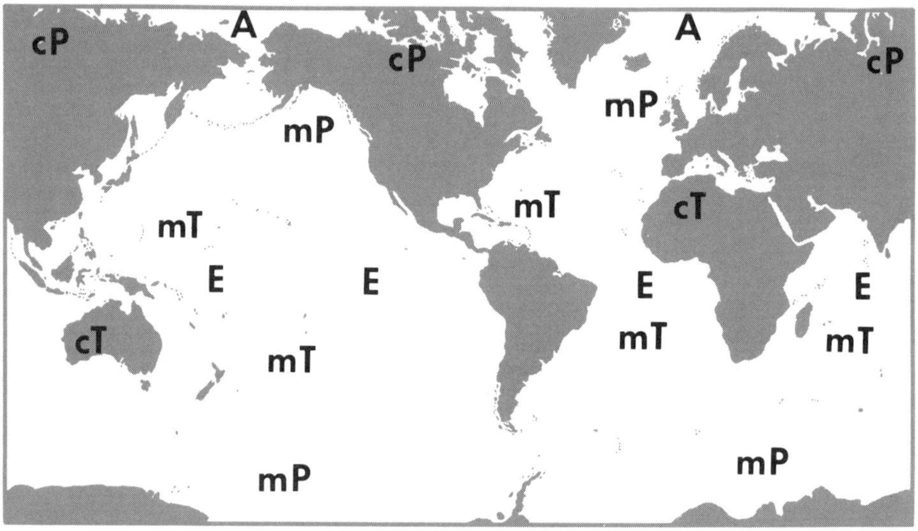

Figure 16-1 Air Mass Source Regions

Note the anomaly in that polar air masses do not originate at the poles. Over the large areas of ice and snow that are typically found at the poles, the Arctic air masses are formed. Because the wintertime in this region is without sun, the air mass is quickly formed. *Polar air* is formed in four regions, the Asian mainland, Canada and Northern U. S., the Gulf of Alaska, and just west of the British Isles. Three regions favor the formation of *Tropical air*. The first two are over the southern regions of the North Pacific and Atlantic, and over the Northern mainland of Africa. *Equatorial masses* are formed only over the water in low latitudes. Since the summertime temperature variation is smaller between land and water, slight modifications occur. The weather is generally less severe than in the winter and the formation of distinct air masses is not so clearly defined.

1602 Air Mass Movement. After a particular air mass leaves its source, its temperature relative to the new surface becomes an additional significant indicator for meteorologists.

An air mass, for example, a *continental polar*, may break out of central Canada and travel southeast to overlie the southeastern United States. If this area had been under a warm air effect, the cold, dry air then would be colder than the contacted surface, fulfilling the requirement to classify it as a cold air mass. The complete symbol for this example has become **cPk,** the **"k"** always being added to any symbol when the mass departs the source region for a warmer surface.

The opposite situation occurs when an air mass, such as a *maritime tropical*, travels over a surface previously cooled by other influences. For example, the **mT** air mass can travel northeast from the Gulf of Mexico earning it the designation of **mTw**; the **w** is the suffix to complete the air mass designation for warm masses.

To affix the **k** and **w**, the air mass must have departed from its source region and must have moved over a surface having a different temperature. The new surface may produce a modification in the air mass, the details of which will be discussed now.

A cold air mass will contain convective activity and turbulence because of the warming of the surface layers. This mixes any atmospheric pollutants and moisture through a deeper layer, producing excellent visibility (outside of clouds). Cumuliform clouds may form and furnish the only significant precipitation. These effects are the result of the unstable nature of the cold air mass.

A warm air mass is very stable by comparison because of the cooling of its low layers. Stratified layers form, inhibiting vertical turbulence. Precipitation will be the slow drizzle type and visibility will be poor. Fog will be prevalent because of surface-layer cooling.

For forecasting purposes air mass motions must be traced. They can be identified by certain conservative properties such as: *potential temperature,** *specific humidity,*† and *dew point*. Other properties such as cloudiness, ambient temperatures, and pressure are too changeable to assist in proper identification and therefore are not used.

1603 Air Mass Modification. Many factors can modify the air mass in its transit: the type of surface it traverses, the topographic features of the surface, and the water content modified by precipitation. The source region letter designation remains, however, until the mass comes to rest again in another source region and the process repeats itself.

* Potential temperature is the temperature an air parcel would have if it were raised or lowered adiabatically to a pressure of 1000 mb.

† Specific humidity is the number of grams of water vapor contained in one gram of moist air.

1604 General Characteristics of Fronts. A *front* represents a sharp discontinuity or transition in weather elements between two adjacent air masses of different densities. Temperature, humidity, wind direction, and other parameters are different on either side of the line of discontinuity. The degree of difference is dependent upon the so-called "strength of the front."

One must remember that the contiguous air masses have vertical extent and thus create a frontal surface. This surface is never vertical but can be thought of as almost horizontal since the incline of the frontal surface is very shallow. When this three-dimensional concept is considered, the isobar shape and the frontal weather appearing on weather charts will then have more meaning.

Medium- to small-scale *frontogenesis*, or front formation, can take place when two different masses are brought together by an outside circulation as shown in Figure 16-2. *Frontolysis*, the opposite process, is the dissolving of a front when the air masses become mixed and indistinguishable.

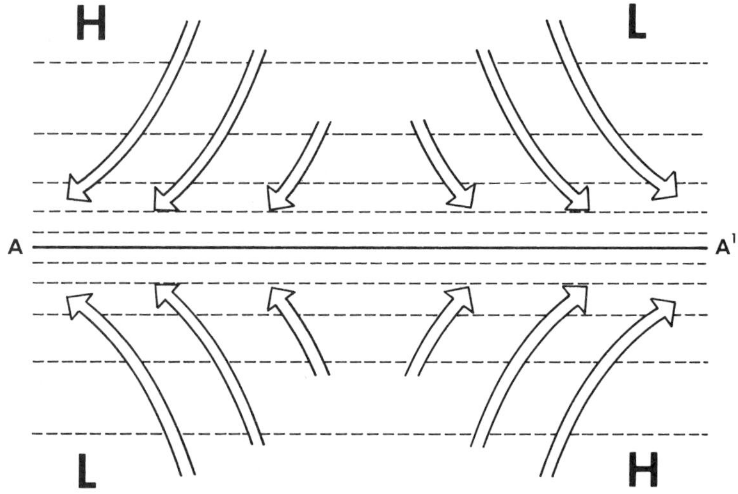

Figure 16-2 Deformation Field: Front Formation Along AA′ Axis (Dotted Lines are Isotherms and Arrows Indicate Wind Direction)

1605 The Warm Front. The warm front exists when a warm air mass is displacing a cold air mass all along the frontal surface or transitional zone. The warm air having a lesser density overrides the cold air mass. This lifting causes adiabatic cooling to occur over a fairly wide horizontal area ahead of the ground-level transition zone. It has an average slope of 1/100 to 1/300, which allows the first cirrus clouds to

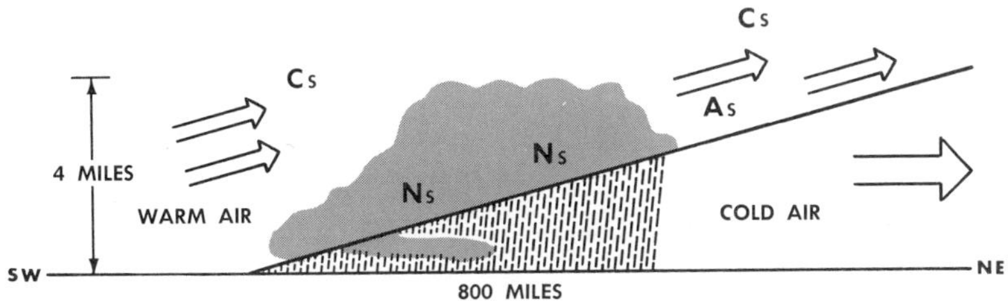

Figure 16-3 Warm Front

appear 700–1,000 miles ahead of the ground-level front, and has an average ground speed of 10 to 15 knots. The rainy area is fairly broad and the rain-producing clouds are generally of the stratiform type in Figure 16-3. Frontal fog may form in the cold air below the frontal surface.

1606 The Cold Front. The opposite condition, a cold air mass displacing warm air, creates a cold front. The cold air with its higher density pushes under the warmer air forcing a steeper slope than that occurring with the warm front; the values average from 1/50 to 1/150. The steeper slope of a cold front is partially due to surface friction holding back the lower levels of the invading air mass. With a warm front, on the other hand, the slope of the retreating cold air mass will be decreased by surface friction.

The frontal weather associated with a cold front lies in a comparatively narrow band and is violently active. Along a well-marked front, cumulonimbus clouds will produce heavy rains, gusty winds, and occasional thunderstorms. The time span required for its passage from the beginning of the weather band to the cold air clearing is generally shorter than that required for a warm front passage. The average ground speed of the cold front is 20 to 25 knots.

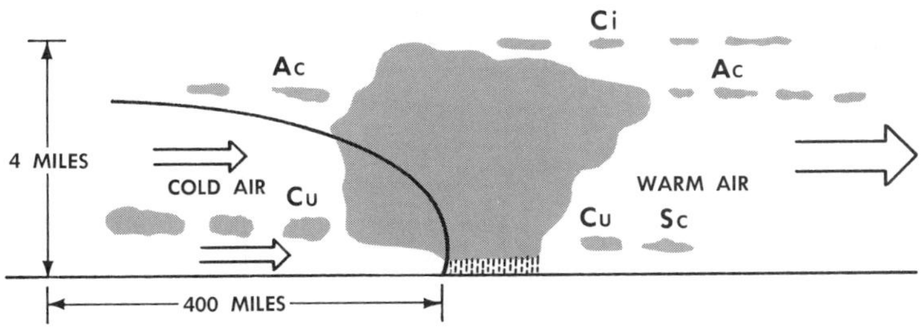

Figure 16-4 Cold Front

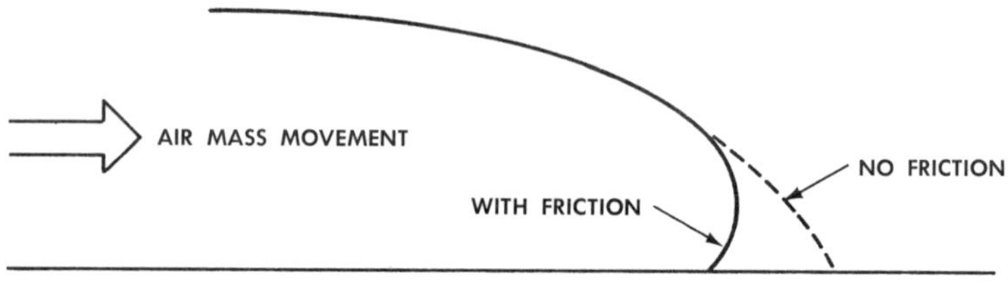

Figure 16-5 Friction Effect

With the exception of cumuloform clouds, the cold air mass will have excellent visibility as a result of mixing in the lower layers from surface warming. By the second night after a cold frontal passage the ground will become completely cooled and the minimum temperature will usually be observed.

1607 The Stationary Front. When there is no motion of the front, the transition zone between the two air masses is known as a stationary front. The weather conditions along the stationary front are similar to those of the warm front. Should the front remain stationary for several days it could become diffuse and frontolysis could occur.

1608 The Occluded Front. When a cold front overtakes a warm front, the merging air mass creates what is known as an occluded front. Since the cold front travels faster than the warm front, the sequence of the involved air masses is cold, warm and cool from west to east (Figure 16-6). When the two fronts are joined, the middle, warm air section is forced aloft.

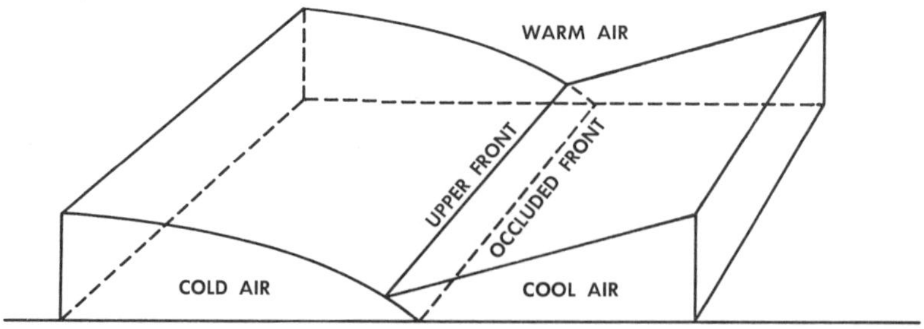

Figure 16-6 Three-Dimensional Occlusion

Since the overtaking cold air mass is colder than the forward cool mass, the result is a *cold front occlusion.* The opposite case is called a *warm front occlusion.* The two types are shown in Figure 16-7. The weather of the occluded front, regardless of the type, resembles that of a simple warm front. The occluding process will eventually cause mixing of the air, thus ending the entire weather system which propagated the frontal movements.

There are upper fronts associated with the occluding process. They are the *upper warm front* in the cold front occlusion, and the *upper cold front* in the warm front occlusion. The front which is forced aloft continues as the upper front. The upper fronts are shown in Figure 16-7. Occlusions provide the greatest hazard for pilots since cumulonimbus clouds (cold front) and their associated turbulences are normally imbedded in the stratus cloud deck (warm front). This latent danger gives no forewarning, by visual means, to a pilot flying in the stratus.

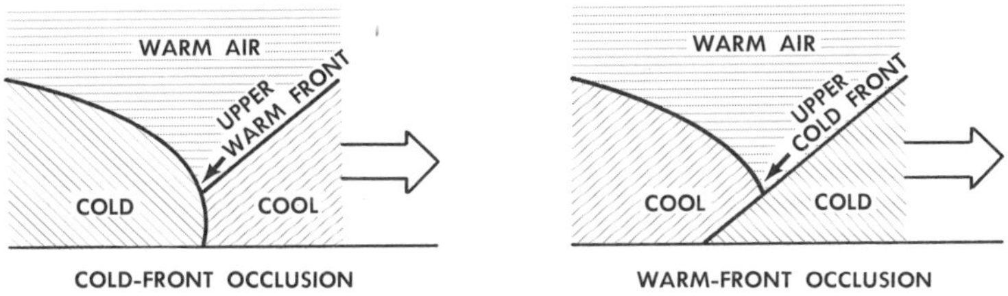

Figure 16-7 Cold- and Warm-Front Occlusions

1609 Thunderstorms. Thunderstorms are classed in two general categories: the *air mass type* and the *frontal type.* In any event, they present one of the most serious hazards the pilot can encounter. The gusts experienced in these towering cumulonimbus clouds have been known to do severe damage to the airframe of a plane that dared to penetrate them. In addition, the rapid changes of static and dynamic pressures leave the pilot with grossly unreliable readings on instruments such as rate-of-climb, altimeters, airspeed. The whole spectrum of weather phenomena including hail, snow, thunder, lightning, torrential rain, and icing may also be encountered. Studies have shown that the most suitable altitude for penetration of these storms is in the lowest third of their vertical extent, but circumnavigation is always the best policy whenever possible. This can usually be done in the case of air mass thunderstorms which are more isolated or separated than the frontal types.

As an aside, the electrical discharges in these storms also present a hazard to aircraft, for they may disturb normal navigational and com-

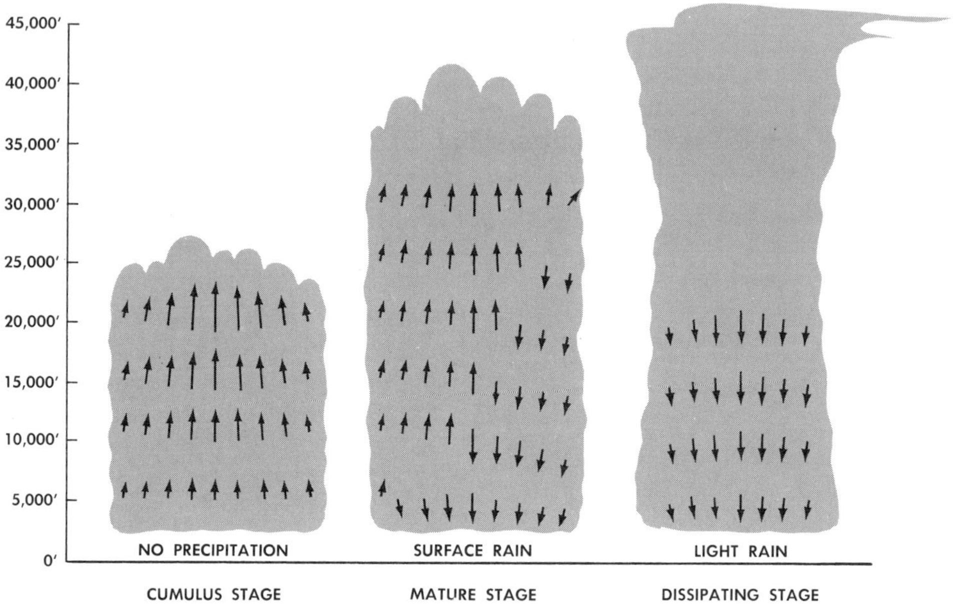

Figure 16-8 Three Stages of a Thunderstorm Cell

munications equipment. Certain types of aircraft direction finders will "home in" on the electrical charge, rendering the information not only useless, but also misleading. The crackle of static in a radio receiver, although not conducive to clarity, is a good clue to the presence of a thunderstorm in the vicinity.

Thunderstorm activity occurs wherever intense convection caused by surface heating or wind-induced surface convergence is present. The upward air motion allows adiabatic cooling and rapid condensation whenever sufficient water vapor is entrained.*

The storm itself occurs as a vertical wind cell or group of such cells. Each cell has a history which includes three basic stages—the *cumulus stage*, the *mature stage*, and the *dissipating stage*.

The *cumulus stage* as shown in Figure 16-8 is characterized by updrafts present at all levels and regions of the storms. Convergence feeds air into the updraft from the surface and from the unsaturated environment of higher levels. Note that during this period, hydrometeors (liquid or solid water particles in the atmosphere) never reach the ground because of the updraft; they grow by coalescence (see Chapter 15) to reach a maximum size at the end of the cumulus stage.

* *Entrainment*—occurs when the surrounding air is drawn into a previously existing air current and becomes a part of it.

The *mature stage* comes into being when there is a combination of an updraft and a downdraft. The downdraft is the result of hydrometeors pulling the air toward the earth. As their size increases, the upsurging air can no longer support them. As the rain or hail falls, it effectively reverses the draft direction in part of the cell. The strongest rainfall then occurs two or three minutes after the first measurable moisture reaches the ground. The outflowing air which spreads away from the downdraft area is much cooler than the surrounding air mass and resembles a temporary cold front.

The *dissipating stage* develops as the updraft ceases; the entire cell is composed of falling air and hydrometeors colder than the surrounding air. The cell dries as the available moisture becomes negligible; the air motions become weak or calm and the cloud disperses. The entire process continues for an hour on the average.

Actual thunderstorms usually consist of two or more cells, each in a different stage of development. It is thought that the surface spreading of cold air from the downdraft sector causes sufficient uplift in an adjacent region to allow self-sustaining convection and spontaneous cell creation. The process continues until the general conditions which created the instability cease to dominate.

1610 Closed, Medium-Scale Pressure Systems. In the atmosphere, closed low-pressure areas are called *cyclones* and closed high-pressure areas are called *anticyclones*. The pressure gradient (isobar spacing) will indicate the wind intensity which may be expected in these closed systems. Cyclones may produce especially strong winds; however, this should not be interpreted to mean all cyclones are to be treated as potentially dangerous storms.

As previously described, the cyclonic system has counterclockwise air motion (in the Northern Hemisphere) because of the coriolis effect. The cyclone which has its associated wind and pressure characteristics identifiable at all levels in the troposphere has a cold central region throughout its vertical extent. This system is called a *dynamic low* and is the result of an intricate interaction of the atmospheric layers. Extratropical storms are primary examples of a cold-core, or dynamic, low.

The other type of cyclone is the *warm-core* (*thermal*) *low*. It is recognized by the fact that it exists only in the lower layers with the vertical pressure gradient becoming anticyclonic in the upper troposphere. Diurnal surface heating, tending to decrease atmospheric density, will produce the most intense low in the afternoon. A typical example of the nondynamic warm-core low is the thermal low frequently found over the Mexican desert plateau.

Anticyclonic systems also have warm- and cold-core divisions. The

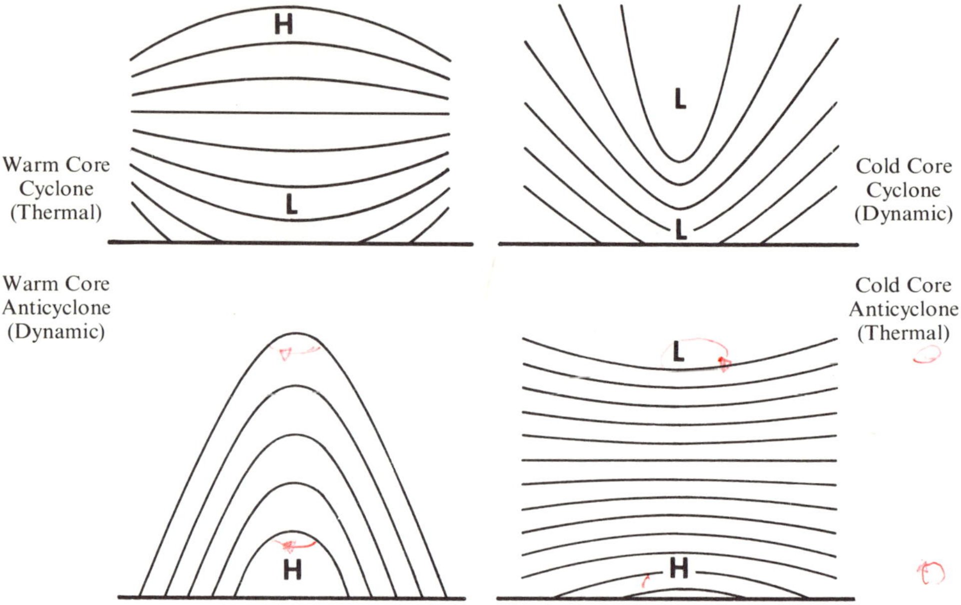

Figure 16-9 Vertical Sections Showing Isobaric Surfaces

warm-core (*dynamic*) *high* has continuous anticyclonic air motion at all tropospheric levels; it is part of the three-dimensional general circulation in concert with cold-core cyclones. In contrast, the *cold-core* (*thermal*) *anticyclone* does not have vertical continuity and is caused by local surface cooling. Figure 16-9 shows the vertical pressure gradients of the four basic systems.

1611 The Life Cycle of an Extratropical Cyclone. Many synoptic observations taken simultaneously provided the evidence needed for research meteorologists to derive the polar-front theory of extratropical cyclones. The storms have continuity and can have predictable movements. The following discussion will limit sequence occurrences to the tropospheric layers; this should not be construed to mean that there are no contributing interactions from the stratosphere.

Exhaustive investigations have indicated extratropical storms can occur along several persistent middle-latitude fronts. The fronts, of course, stand in separation of two unlike air masses of different densities and usually between the westerlies and upper-latitude easterlies. With opposite winds and temperatures on both sides of the front a wave or perturbation can easily appear. These waves are thought to be caused by outside influences such as irregular terrain, other storms, and changing frictional components at water-land boundaries. The perturbation is classified as either "stable" or "unstable." A stable wave will disappear or be damped out while an unstable wave will develop into lasting cyclonic disturbances.

Storm development is called *cyclogenesis*. Various weather agencies are constantly trying to forecast this process by investigation of all favorable and unfavorable indications in both time and space.

The wavelike indentation (Figure 16-10A) will move with the winds creating a leading warm front and a following cold front. The warm air mass is caught up in the space now referred to as the warm sector. The

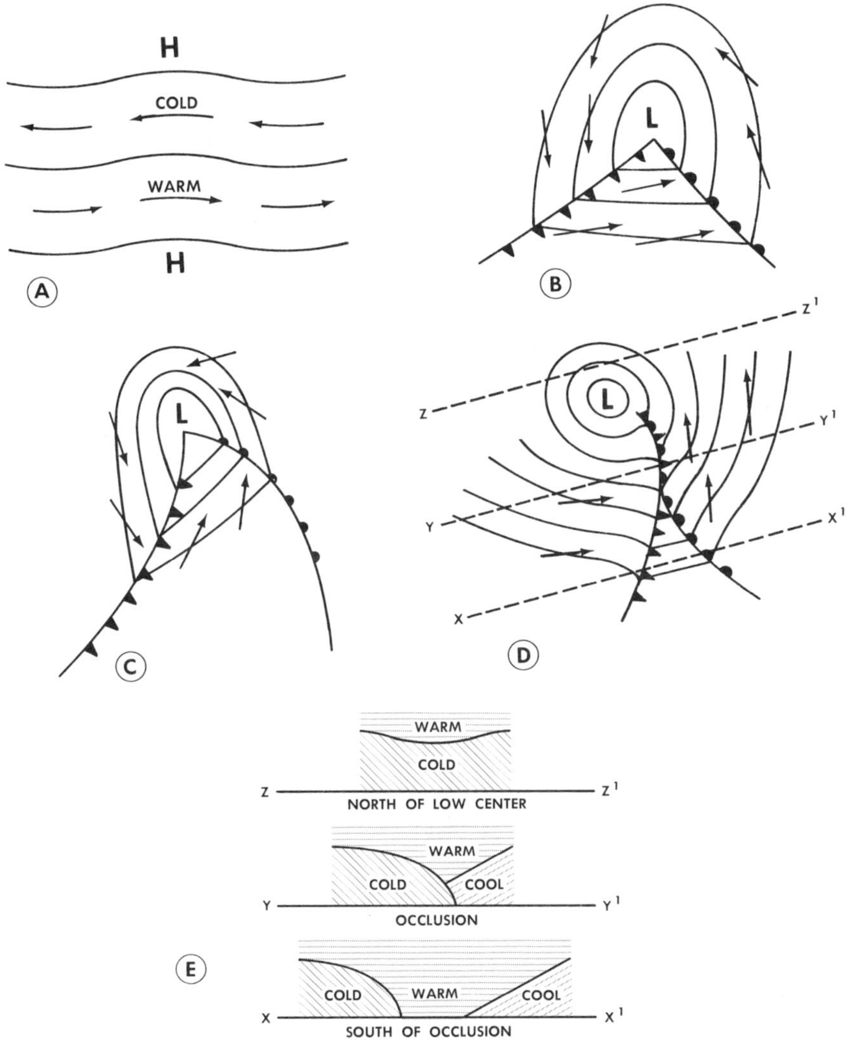

Figure 16-10 Life Cycle of Extratropical Cyclone

low appears at the apex of the wave and is accompanied by cyclonic circulation (Figure 16-10B). Because there is wind shear in the proper direction, the air is moved against the distorted front, permitting continual growth (Figure 16-10C).

The manner in which the cold front overtakes the warm front is shown in Figure 16-10D. At this stage the cyclone has reached its peak of intensity and is considered to have reached "maturity." The occluding process now begins pushing the warm air aloft.

During the building process the entire storm system will have moved in an easterly direction in either hemisphere. The isobar direction in the warm sector can be a reliable clue to the storm's future movement. If the polar front lies beneath the jet stream, it will have a strong steering effect on the storm system. However, the cyclone's movement begins to slow and may even reverse its direction slightly as it dies.

The storm center fills (i.e., the pressure difference decreases from the edge of the cyclone to its center) as the cold air flows around it. Eventually the two intermingling cold air masses of the occluded front will form one homogeneous mass and the cyclone will cease to exist.

The isobars at the front always kink away from the low. Synoptic pressure observations have verified that a small surface pressure trough extends from the low along the warm and cold fronts making the isobars shift sharply at the fronts. In other words, there will be a minimum value of atmospheric pressure experienced as a front passes.

Cross sections showing the warm sector, occlusions, and cold air poleward of the low center are shown in Figure 16-10E.

16-11 Isobaric Pattern of Idealized Cyclone Family

1612 Cyclone Families. In many cases, extratropical cyclonic waves develop in succession along the polar front westward from the original storm. Very often a series of cyclones occur rather than one by itself. Some of the later waves may be stable and disappear almost immediately after first appearing. A representation is shown in Figure 16-11. Note that the various stages depicted represent a time and space sequence similar to Figure 16-10.

Additional Reading

Anon., *Meteorology for Mariners*, Her Majesty's Stationery Office, London, 1956.

Barry, R. G., and Chorley, R. J., *Atmosphere, Weather, and Climate*, Holt, Rinehart, & Winston, 1970.

Byers, Horace R., *General Meteorology*, 3rd ed., McGraw-Hill, Inc., 1959.

Donn, William L., *Meteorology*, 3rd ed., McGraw-Hill, Inc., 1965.

Petterson, Svere, *Weather Analysis and Forecasting*, 2 vols., McGraw-Hill, Inc., 1956.

Riehl, Herbert, *Introduction to the Atmosphere*, McGraw-Hill, 1965.

Taylor, George F., *Elementary Meteorology*, Prentice-Hall, Inc., 1954.

CHAPTER SEVENTEEN

Tropical Cyclones: Hurricanes and Typhoons

1701 Introduction. Many times since men have been venturing forth to face the elements, tropical cyclones have caused ships at sea to founder. Why do these storms form? How do they exist? How do sailors avoid such catastrophic storms? What actions should be taken when ships are caught in such storms? Questions such as these will be answered in the following pages.

Known as *hurricanes* in the Atlantic Ocean and as *typhoons* (Chinese, *ta-feng*, great wind) in the North Pacific Ocean, tropical cyclones have been dubbed with other special names such as *willy-willies* off the coast of Australia.

These storms are noted for their extreme winds—in some cases in excess of 200 knots—heavy rains, and abnormally low pressures. By international agreement, the minimum wind is 65 knots for a storm to be classed as a hurricane or typhoon.

Hurricane Hilda in center of Gulf of Mexico on October 1, 1964, photographed by Tiros VII

1702 Characteristics of Tropical Storms. When compared to extratropical storms, tropical storms are usually small, circularly shaped, intense low-pressure zones. Very steep pressure gradients exist from the periphery to the center. These storms have no fronts and are contained in a single air mass. The pressure drop from the periphery to the eye is normally from 20 to 70 millibars, with larger drops not uncommon. The lowest pressure ever recorded was 26.185 inches (887 mb), in the eye, near Luzon, P. I., on August 18, 1927. The pressure gradient is most intense near the center; this intensity causes the increasingly high winds in the inner area.

The winds are in a constant circular counterclockwise motion in the Northern Hemisphere, but the center of the storm is a region of calm. A ship or structure in the direct path of the cyclone will first experience the wind building gradually from one direction, then a calm, and finally, a sudden reversal with only a short interval for the acceleration. This change in direction is usually the cause of destruction because of its sudden onset.

The *eye* of the storm, ten miles across on the average, is the center of low pressure. This central area is generally clear because of the warm, dry nature of the subsiding air necessarily present for the storm to exist. Surrounding the eye is the *wall* consisting of cumulonimbus clouds which indicate the extreme instability or rising motion in this area. Heavy rains fall in the wall area immediately adjacent to the eye; 6 to 12 inches of rain in 24 hours are not uncommon, and rainfalls of several feet have been recorded.

The cloud structure out from the central zone is generally uniform in all directions. Altostratus (As) and nimbostratus mix to form a solid cloud layer to a radius of 200 miles in extreme cases. Cirrus (Ci) mixed with cirrostratus (Cs) is high and reaches to the outer regions of the storm.

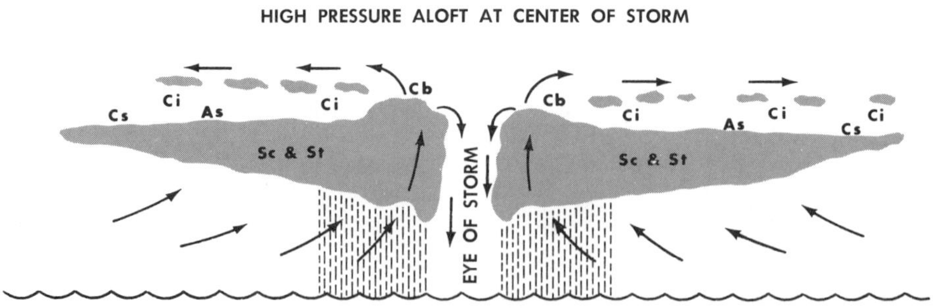

Figure 17-1 Cross Section of a Tropical Cyclone

The tropical cyclone is nature's method of continuing the general water budget of the atmosphere. Since the absolute moisture content of the tropics is highest in the summer, the autumn hurricanes help to release the excess water as the temperatures begin to decline. The released latent heat results in a partial maintenance of air temperature not otherwise possible. New moisture-laden air is continually entrained into the system to continue the water- and heat-releasing processes.

1703 Methods of Formation. A definite identifiable triggering mechanism has not yet been isolated. Many conditions must be present simultaneously, but the actual formation of a tropical cyclone is still considered to be a random occurrence.

One possible mechanism involves an *easterly wave* or trough which develops along the easterly flow and breaks the normal trade-wind temperature inversion. This means that the usually well-defined layer between the moist lower air and dry upper air must be broken on a large scale by vertical motions. The squall line along the wave separates clear skies to the west from stormy, rainy zones behind, in the east.

A vortex may form along this squall line should an upper-level anticyclone happen along. The high pressure aloft will literally pump air up

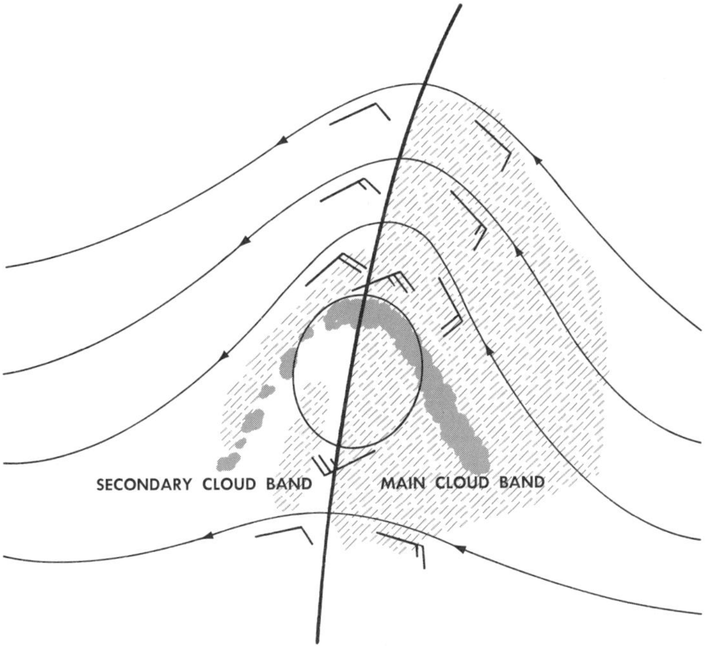

Figure 17-2 Easterly Wave in Northern Hemisphere, as Seen From Above

and outward from the eddy, creating a low pressure at the earth's surface. This process alone may cause inner winds, at speeds of 45 knots, to form closed isobars in an area which normally has little pressure gradient.

At this stage of development, there has been much subjective reasoning concerning a storm's ability to intensify. The central subsiding air must have an extremely high potential temperature. In fact, the Joint Typhoon Weather Central, Guam, often uses upper air potential temperatures as a primary clue to possible storm formations. A typhoon has never formed in an area where the ocean surface temperature has been lower than 26°C, indicating a positive requirement for air-ocean interaction. As the latent heat source is left behind when moving over land, cyclones always lose intensity or may even die. These conditions and many others have been advanced as factors which intensify and prolong tropical storms. Much research is planned to ascertain the origin and control factors of these storms.

1704 Storm Progression. The spawning ground for tropical cyclones is generally in the doldrums. They are rarely observed within 5° of the equator. Thus this absence indicates the imperative effect of coriolis to develop and maintain the cyclonic motion. After the storm has taken form, the normal track in both hemispheres is initially from east to west. As the cyclones build intensity, they become violent and potentially destructive.

The individual storm tracks are radically diverse, but general statements can be presented concerning the average paths. The centers at first move slowly away from the equator initially with the easterly flow. They then develop gently curving paths toward the poles in each hemisphere. This is the beginning of the recurvature process which culminates in an eventual storm path directed toward the northeast in the Northern Hemisphere and southeast in the Southern Hemisphere. The recurvature trend seems to follow the warm surface currents such as the Gulf Stream System, Kuroshio, and East Australian Currents as if to follow unconsciously a potential source of longevity—a heat sump.

The primary meteorological reason for the recurvature is the pull of the westerlies after the affected storm is far enough from the equator. The tropical cyclone forecasting centers are continuously predicting when a particular storm will be influenced by the prevailing flow. Also, the effects of the upper air flow are not to be overlooked. Where the anticyclone goes aloft so goes the surface cyclone. The high aloft is mandatory for the maintenance of the wall updrafts.

These cyclones travel at 10 to 12 knots before recurvature. They accelerate as they turn away from the equator and proceed at speeds up to

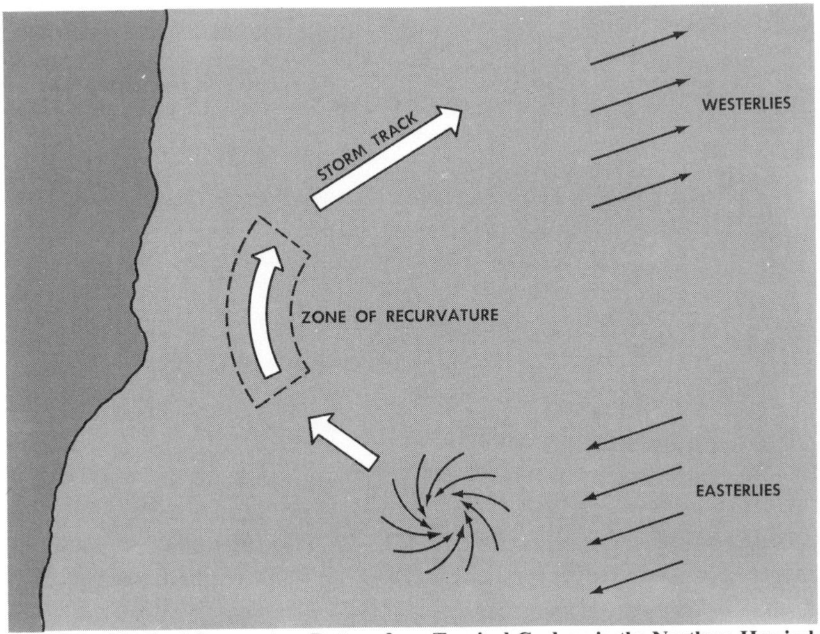

Figure 17-3 Idealized Recurvature Pattern for a Tropical Cyclone in the Northern Hemisphere

30 knots. The storms become sluggish when passing over large land masses, sometimes diffusing. If they remain at sea, they generally become extratropical wave cyclones and blow themselves out as the energy sources (warm water) become unavailable.

One Atlantic hurricane was tracked closely as it passed over the cold and warm tongues associated with the Gulf Stream System. The data clearly demonstrated a lesser wind intensity when the hurricane was over water tongues with surface water temperatures below 26°C than when it was over warmer water areas. The technical report indicated the presence of a very possible interaction.

The frequency of storms in the various localities of the world is shown in Table 17-1. No tropical storms have ever been reported for the South Atlantic or eastern South Pacific Oceans. These two regions are continuously dominated by cold surface currents, the Benguela and Humboldt currents respectively, which very likely maintain subcritical overlying air temperatures thereby preventing tropical storm formations. In contrast, the western North Pacific has the largest storm frequency, followed by the western North Atlantic and Bay of Bengal. Sometimes there is more than one storm active in a particular region at once. They may even join to form a monster storm as was the case in August 1964 just south of Honshu Island, Japan.

1705 Tropical Storm Navigation. While certainly all areas of a tropical cyclone are dangerous, a ship may maneuver to sectors where winds and seas are less intense.

Table 17-1 Monthly Tropical Cyclone Occurrence With High-Speed Winds (TIROS Data Not Included)

	YEARS OF DATA	J	F	M	A	M	J	J	A	S	O	N	D
North Atlantic Ocean	68	-	-	-	-	0.1	0.4	0.5	1.5	2.6	1.9	0.5	-
Eastern North Pacific Ocean	30	-	-	-	-	0.1	0.8	0.7	1.0	1.9	1.0	0.1	-
North Pacific Ocean (West of 170°E)	36	0.4	0.2	0.3	0.4	0.7	1.0	3.2	4.2	4.6	3.2	1.7	1.2
North Indian Ocean (Bay of Bengal)	36	0.1	-	0.2	0.2	0.5	0.6	0.8	0.6	0.7	0.9	1.0	0.4
North Indian Ocean (Arabian Sea)	23	0.1	-	-	0.1	0.2	0.3	0.1	-	0.1	0.2	0.3	0.1
South Indian Ocean (West of 90°E)	70	1.3	1.7	1.2	0.6	0.2	-	-	-	-	-	-	0.1
South Indian Ocean (Off N.W. Australia)	Unknown	0.3	0.2	0.2	0.1	-	-	-	-	-	-	-	0.1
South Pacific Ocean (West of 160°E)	105	0.7	0.4	0.6	0.2	-	-	-	-	-	-	0.1	0.3

Violent storms create heavy seas. The term *sea* refers to the surface conditions which exist within the storm's confines. White caps are visible and the direction of wave travel is completely dependent upon the wind direction. Constantly changing winds create confused seas. *Swell* is composed of relatively smooth waves which are observed away from and ahead of the winds. The seas or wind waves are created in the area referred to as fetch (see section 1907), with a longer fetch producing higher waves for the same wind speed.

For navigation purposes, tropical cyclones are considered as two semicircles. In the Northern Hemisphere the semicircle to the right of the cyclone track is the dangerous semicircle. The direction of the winds coincides with the cyclone itself, when vectorially added. Whereas, on the left, the winds are opposite to the direction of cyclone travel. The right semicircle has a much longer fetch allowing the seas to build to dangerous heights. The duration of wind action on the left will be much shorter in comparison. The windspeed difference between the two semicircles could be as much as 60 knots after recurvature, if the speed of travel of the storm were 30 knots. Note that when the storm direction changes, the semicircles are geographically relocated.

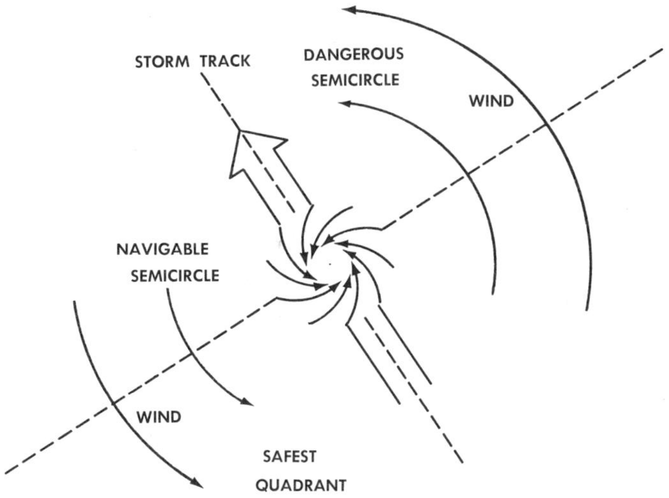

Figure 17-4 Semicircle Diagram for Northern Hemisphere

A ship caught in the right semicircle should bring the wind to the starboard bow in order to evade the storm's path as much as possible. Should the ship be forced to abandon its original track, it should then proceed in order to get as far as possible from the storm center. If in the left or navigable semicircle, the ship should bring the wind to the starboard quarter and run with the seas. This action will take it away from the storm quickly and allow for a relatively safe and smooth ride.

1706 Locating Storms. Weather satellites are proving themselves invaluable in locating cyclone centers. They are finding storms in very early formation stages in remote areas, where they might otherwise go through an entire life cycle undetected.

Routine air reconnaissance by the Weather Bureau, Air Force, and Navy, depending upon the area of designated responsibility, is a very valuable means of cyclone tracking. Valuable data from within storms have been collected and aid in later basic research. Storm seeding, whenever approved, is also done from the aircraft.

Both land- and air-based long-range radars can present excellent pictures of a cyclone's precipitation pattern. Once detected, a storm may be continuously tracked by radar.

The typhoon and hurricane warning systems are well staffed with experienced personnel. The many subjective and objective methods used are becoming more sophisticated and accurate in predicting the cyclone's track and intensity. However, the forecast is most valuable to the person who has a basic environmental understanding necessary to take prompt action based on the information available.

Had the reliable locating and tracking systems now in operation been available during World War II, perhaps the total loss of three destroyers which occurred in one of the two great storms now known as "Halsey's Typhoons" would have been prevented.

Additional Reading

Anon., *Meteorology for Mariners*, Her Majesty's Stationery Office, London, 1956.

Battan, L. J., *The Nature of Violent Storms*, Anchor Books, Doubleday & Company, Inc., 1961.

Byers, Horace R., *General Meteorology*, 3rd ed., McGraw-Hill, Inc., 1959.

Dunn, Gordon E., and Miller, B. I., *Atlantic Hurricanes*, Louisiana State University, 1960.

Harding, Edwin T., and Kotch, William J., *Heavy Weather Guide*, U. S. Naval Institute, Annapolis, Md., 1965.

Malkus, Joane Starr, "The Origin of Hurricanes," *Scientific American*, August, 1957.

Petterson, Svere, *Introduction to Meterology*, McGraw-Hill, Inc., 1958.

———, *Weather Analysis and Forecasting*, 2 vols., McGraw-Hill, Inc., 1956.

Riehl, Herbert, *Tropical Meteorology*, McGraw-Hill, Inc., 1954.

Taylor, George F., *Elementary Meteorology*, Prentice-Hall, Inc., 1954.

CHAPTER EIGHTEEN

Synoptic Meteorology: A Panorama

1801 Introduction Each one of us, at one time or another, has taken on the role of forecaster, to some degree. A practiced eye is usually cast on the state of the sky by the man who is about to wash his car, or the woman who is about to hang out the wash. Forecasting is big business, and the weather outcome of almost any major outdoor sporting event can be insured through some underwriter. History is replete with examples of the importance of timely environmental data. The invasion of Normandy by the Allies in World War II was successful in a major degree because the Axis powers were convinced that the weather would prevent such a move.

One of the first attempts to apply scientific or systematic techniques to foretelling weather arose as the result of a naval disaster. In the middle of the nineteenth century, the French, in league with the British, attempted to blockade the Russian port of Sevastopol. An un-

Courtesy: Michael Airey Griffin, Berkeley, California

expected windstorm caused the sinking of a French capital ship, which so upset the French Minister of War that he commissioned a prominent astronomer to devise a method of foretelling the capricious weather that causes such disasters. Prior to this time, the variability of the weather had been attributed to circumstances, and intelligent analysis was hampered by the inability to communicate over long distances. It was Benjamin Franklin's inventive mind that prompted the discovery that the same weather probably moved from one area of the earth's surface to another. The advent of the telegraph allowed the instantaneous transmission of warnings of approaching weather. Thus the first meager efforts begun by the French have since grown to a global network of international cooperation.

1802 The Basic Process. There are many ways to forecast the weather but the one that is practiced by the majority of scientific meteorologists is called *synoptic meteorology*. There are other methods of making a forecast which range from the ouija board, through the farmer's almanac, to *weather typing*.* All of these are based upon different principles, some more scientific than others. The term synoptic is a derivative of the Greek *syn*-together + *opsis*-a sight, since this quite accurately describes the function of this type of observation. It is intended that an instantaneous picture be taken of the atmosphere at preselected times, in an effort to effectively interpret the forces at work on the air masses, and the boundaries between them. This is literally designed to give the analyst the *panorama*, the skeleton around which the structure of professional meteorology is built in the Navy, the other armed services, and the National Oceanic and Atmospheric Administration (National Weather Service). The World Meteorological Organization, now under the auspices of the United Nations, serves as the instrument for international effort. Many facets of weather reporting are consolidated under their aegis. Attention will be focused on only a few significant parts of this reporting system.

The important aspects of synoptic meteorology are broken into the following four steps:

Observing and reporting the weather
Collecting and displaying the data
Analyzing the information
Interpreting and forecasting the weather.

The discussion will be initially confined to the data for the surface conditions.

* Weather typing is an attempt to identify the weather pattern of today with that of some time in the past. It is assumed that the same weather which followed the historical pattern will occur tomorrow.

1803 Observing and Reporting. Through international agreement the code for reporting weather observations has been reduced to a numerical representation. Every observer throughout the world makes his weather observations at the same four times during the day, on a schedule predicated upon the time at Greenwich, England, or Greenwich civil time. Greenwich time can always be identified by the letter "Z" behind the four-digit number, for example, 0800Z means 8:00 a.m. at Greenwich but 3:00 a.m. (0300) eastern standard time in New York. At 0000Z, 0600Z, 1200Z, and 1800Z each day, these global observers take simultaneous readings and observations of temperature, clouds, humidity, wind, pressure, visibility, and precipitation. These "weather elements," as they are called, are immediately reported as a series of numbers which are transmitted on teletype. Now, at 1200Z for example, at any global location, a teletype machine can receive the reports from all over the world. They would look something like this:

491 83214 24516 24706 66222 05328 74546

This may seem to be a meager bit of information but, translated, it says:

> The skies are overcast at Monterey, California, with stratus clouds whose bases are between 300 and 600 feet. In addition to these low clouds there are also thick altostratus and dense cirrus. In the past six hours there has been rain. It began about three or four hours ago and during this time, 0.45 inches has accumulated. The wind is from the northwest at about 13–17 knots. The temperature is 6°C* and the dew point is 5°C. At present the surface visibility is reduced to $1\frac{1}{2}$ miles in continuous slight drizzle. The atmospheric pressure is 1,024.7 mbs and has changed a total of 2.8 mbs in the past three hours. This change initially was a fall but has since risen and the pressure is greater than it was three hours ago.

1804 Collection and Display. Indeed, the numerical code tells a comprehensive story. Within a short time the meteorologist can copy all the reports from the United States and his assistants can begin plotting them in the proper locations on a weather chart. The *station model* is the accepted standard format for translating the above numerical groups into a more readable arrangement. This information, as given above, plotted on a surface map in station model form is shown in Figure 18-1. The maps on which these station models are plotted have

* At this writing, the United States is displaying Fahrenheit temperatures in most reports, for public convenience.

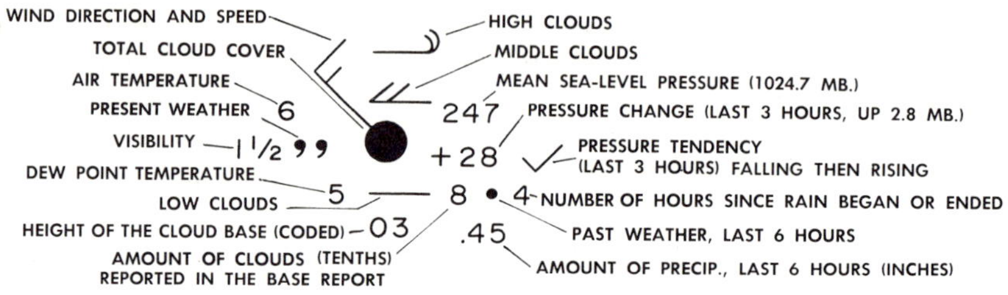

Figure 18-1 Land Station Model

the small basic circle and the number of the station already affixed. Only the coded data must be taken from the teletype and reproduced on the map. When this is completed, the meteorologist has the data available to produce the panorama that is necessary to formulate a forecast.

If the area to be examined contains ocean areas, the reports from ships will differ slightly. As one would imagine, certain supplementary information is necessary and desirable from a moving ocean platform. First, the latitude and longitude of the ship's position must be given, so that the data may be plotted in the correct location. In addition to the regular land station data, the ship includes information on the sea surface temperature, any observed sea ice, period and height of waves and swell, direction of swell, plus the ship's course and speed made good

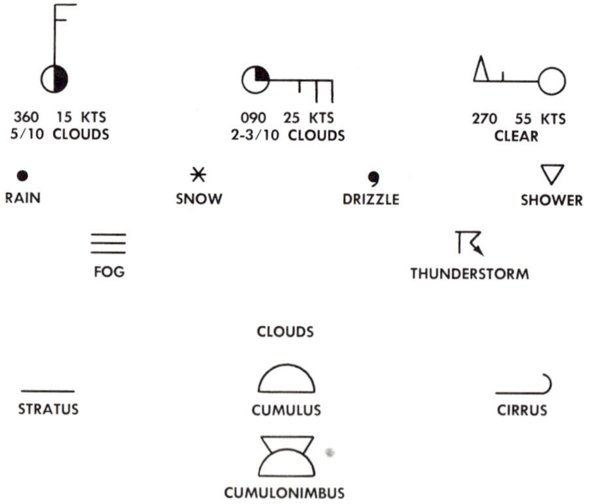

Figure 18-2 Some Common Station Model Symbols

during the last three hours. The course and speed are vital since the barometric tendency has little meaning unless it can be determined that the change of the atmospheric pressure is due to a change of the air mass, or the motion of the ship itself, in crossing the pressure gradient. Figure 18-3 is an example of a ship station model.

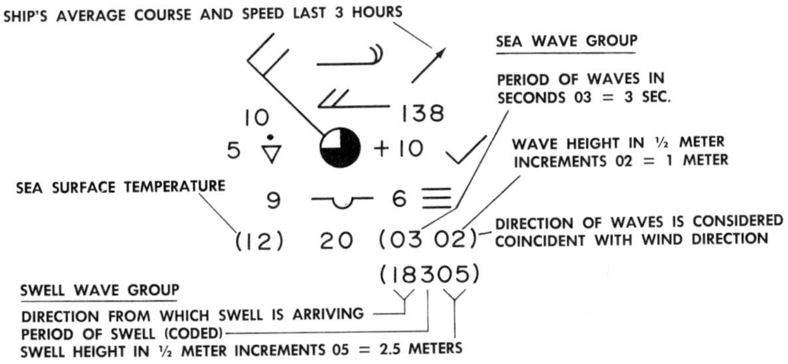

Figure 18-3 Ship Station Model

1805 Analysis. The next step in synoptic meteorology is the analysis of the data which has been plotted. It is not the purpose of this text to present all of the details of analysis but some of the principal steps in the method are desirable for an understanding of the process. An experienced analyst will never approach the latest chart without consulting the previous charts, since continuity in weather movement is an often observed characteristic. In fact, it may be standard procedure for the aerographer, even before beginning the plot, to delineate on the chart the previous three or four positions of the following features:

Highs and Lows	Trough and Ridge Lines
Frontal Positions	Frontolysis

This is one of the most important steps in the analysis process. It is fundamental that synoptic meteorology be concerned with the *movement and interaction of particular air masses.* Reasonable air mass movement, consistent with known behavior, is a requirement for an intelligent analysis.

The next step is the preliminary location of the fronts. A number of weather elements dictate the frontal locations. Usually, a glance in the area of anticipated movement, as shown by the past performance of the front, will reveal definite shifts in the wind direction, a temperature

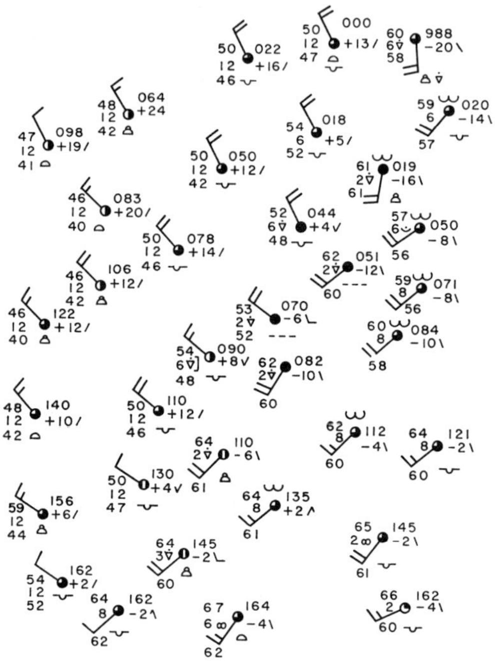

Figure 18-4 Segment of Unanalyzed Chart Showing Station Model

contrast between stations on either side of the front (the dew point temperature is a particularly good indicator since the moisture content of an air mass remains fairly stable) and/or a readily identifiable change in the pressure tendency. As an aid in this process of frontal identification, and to make the chart more graphic, for example, for briefing pilots, many analysts will lightly tint stations with significant weather using the following color scheme:

Precipitation	green
Fog or restrictions to visibility	yellow
Thunderstorms	red

Once this step is completed, the areas of frontal location often stand out and are easily delineated.

By placing a sheet of clear acetate over the plotted chart, the analyst can sketch in the isobars with a grease pencil, yet allow himself latitude to alter the isobaric pattern as the analysis progresses. When the job is finished, the acetate is removed and placed under the chart in the same relative position. By illuminating both from below, the finished analysis can be traced onto the plotted chart.

In examining the pressure pattern the initial move is to locate the

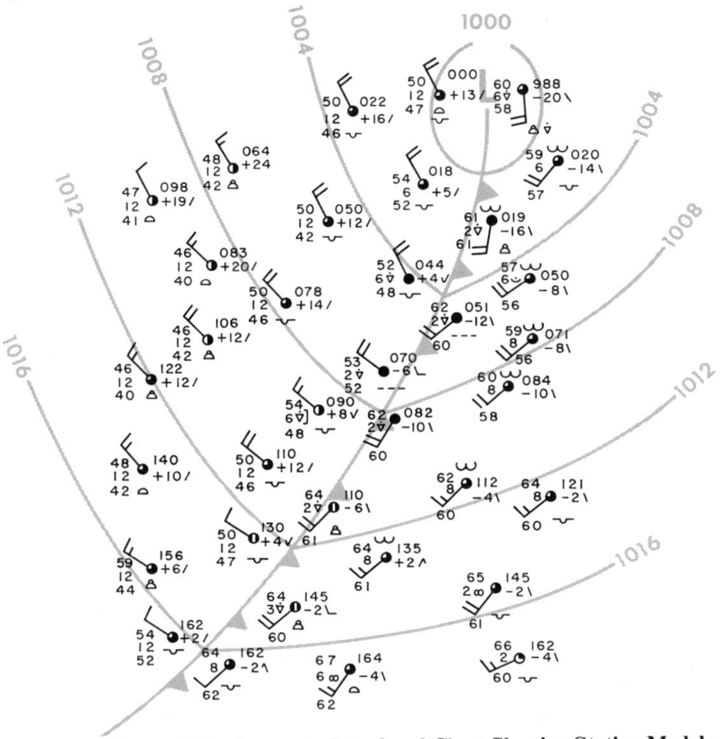

Figure 18-5 Segment of Analyzed Chart Showing Station Models

highest and lowest pressure values in the plotted data for a particular area. Since the isobar is a line of equal pressure, *isobars will never touch, cross one another, or split*. Ideally, on a global basis every isobar would be continuous and closed.

The isobars will also show an agreement with the wind reports. Recalling the principles laid down in Chapter 9, notice that the wind tends to be turned inward towards the center of a low and outward from the center of a high due to the frictional effect of the surface. This effect will be more marked over land (up to 45°) than over water (about 10°). The intensity of the wind must also show some agreement with the spacing of the isobars. The higher the wind speed, the greater must be the pressure gradient responsible, consequently the closer will be the spacing of the isobars. The resulting pattern should be smooth with consistency of spacing between adjacent lines. Some meteorologists insist that there is an element of an art form as well as strict scientific objectivity in pressure analysis. It is true that any departure from the field of pressure should be closely examined and evaluated for possible transmission and plotting errors. A glance at the previous report of a particular station may shed some light on the credibility of the report in question.

As the isobar crosses the frontal line, a sharp change of direction

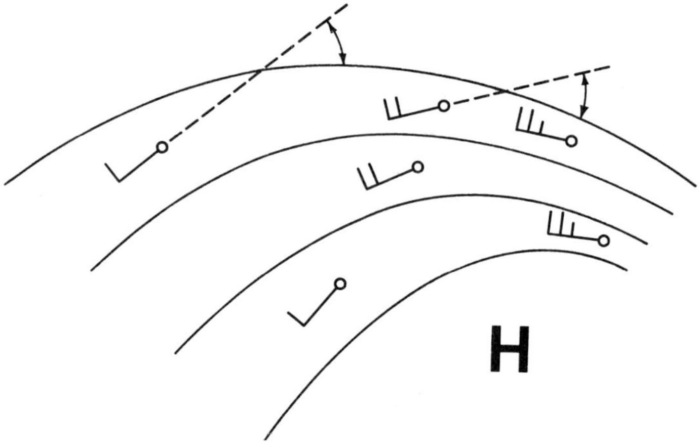

Figure 18-6 Winds and Isobars

should take place. This "kink" or V-shape of the isobar emphasizes the discontinuous nature of the intersection of the air mass boundary with the surface and should always point toward an area of higher pressure. The basic interval of pressure analysis is 4 mbs, above 20° latitude. Closer to the equator the interval should be 2 mbs, reflecting the lesser variability of pressure within this region. By counting from 1,000 mbs, the proper selection of isobars will result.

Other analyses can be done on the surface chart. The analysis of barometric pressure change, for example, is represented by a family of lines called *isallobars*. An analysis of the dew point produces a family of *isodrosotherms*. The most familiar supplementary analysis associated with the pressure analysis is the temperature field. These isotherms are plotted every five degrees, starting with 0°C and are drawn in red.

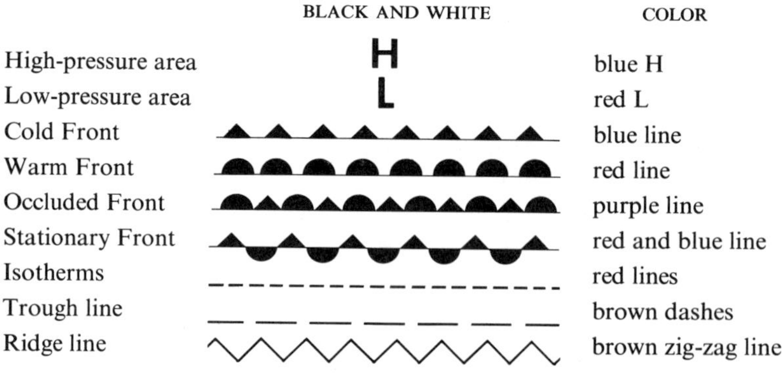

Figure 18-7 Line Representations for Weather Charts

Figure 18-8 Preparing a Sample Analysis

Since charts are produced in both black and white (those transmitted over the facsimile system) and in color, two representations are used as indicated in Figure 18-7.

The *surface map* is invaluable to the meteorologist. It is also more familiar to the majority of laymen than any other analysis product and can be obtained by several methods. A completely analyzed chart is broadcast on the facsimile network and can be copied by any agency or office that has the equipment to do so. The *facsimile process* is a method by which the signal is transmitted by telephone lines or by radio at sea. At a remote location a scanning scribe progressively burns the image of the transmitted map on special paper in the receiving equipment.

In addition, the practicing meteorologist may also analyze the significant area of the chart which concerns his local operations, particularly if he is called upon to give specific service for a small area (e.g. will it be foggy here at 0800 tomorrow?).

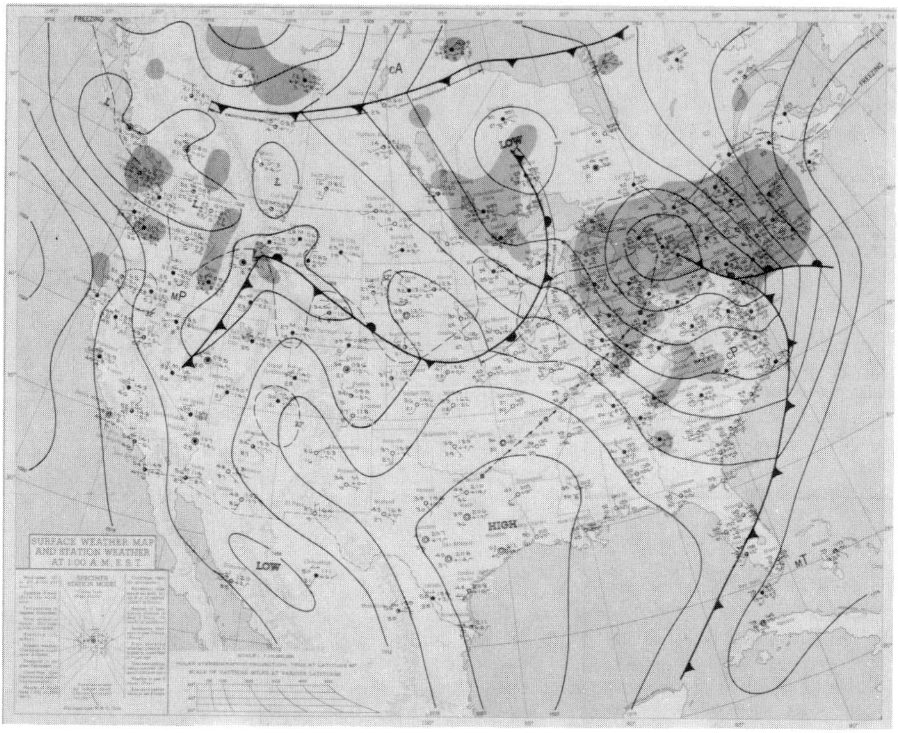

Figure 18-9 Surface Synoptic Chart

Another extremely useful tool of the meteorologist is the *Airways Sequence Report*. It is transmitted hourly on the teletype, as part of data from a group of stations from a particular geographical area. It is a very short report of the state of the weather at selected airports and looks like this:

TUL M8 ⊕ 3F 115/56/55/0213/985

This particular one is translated:

"Tulsa—the ceiling is measured at 800 feet overcast, visibility 3 miles in fog. Pressure 1011.5 mbs, temperature 56°F, dew point 55°F, wind from 020° at 13 knots, and the altimeter setting is 29.85 inches of mercury."

It can be clearly seen that the report is broken into two distinct groups, the sky and visibility group, and the temperature-pressure-wind group. Note the ceiling group is found after the three letter station identifier. It is usually preceded by a letter telling about the ceiling or how the ceiling observation was made such as: M for measured, E for estimated, U for unknown. The next number is the height of the clouds in hundreds of feet, followed by the symbol which gives the intensity of the cloud cover; O for clear, ⦶ for scattered (0.1 to 0.5), ⦷ for broken (0.6 to 0.9), and ⊕ for overcast (ten-tenths cloud cover). The difference between *scattered* and *broken* has a great deal of significance to the aviator, since a broken sky is defined as a ceiling and a scattered sky is not. The height of any ceiling in conjunction with the surface visibility determines the rules he must follow: *Instrument Flight Rules* (IFR) are quite demanding while *Visual Flight Rules* (VFR) are much less restrictive. Finally, in this group is found the visibility expressed in miles, and a code indicating the present weather. Some symbols are: S for snow, R for rain, K for smoke, H for haze, F for fog.

The last group represents five parameters separated by the slant sign (/). In order they are:

sea-level pressure / temperature / dew point / winds / altimeter setting

As in the station model, only the last three digits of the sea-level pressure are reported. The 10, of 1011.5 mbs, as in the example, is not reported or if the pressure were 986.3, the 9 would be omitted. The wind is reported in four digits, the first two indicate the tens and hundreds digits of the direction from which wind blows, on a 360-degree basis. The two digits following represent the speed in knots. Therefore, 0213 indicates the wind is from 020° at 13 knots. (The altimeter setting is used by pilots only, to adjust their cockpit instruments.)

The difference between the professional meteorologist and the amateur can be identified in many ways. One of the most significant differences is that the amateur looks only at the two-dimensional picture of the weather, either in time or space, but the professional considers the state of the atmosphere from three linear dimensions, with the fourth dimension of time included. The professional is vitally interested in all the various aspects of the behavior of the atmosphere and, in addition, its interaction with the sea.

In pursuit of this three-dimensional space outlook, the most popular upper-air chart is the *500 mb chart*. Since this represents the boundary between halves of mass of the atmosphere, it is significant. In addition, although the ceilings of aircraft flight are constantly rising, many, many aircraft travel at or near this level. These constant-pressure charts are analyzed much the same as are surface charts with the exception that the isobar is replaced with the contour line or *isohypse* (Greek, *iso*-same and *hypsos*-height), the altitude at which the pressure of interest occurs. Upper-air charts are drawn at 1000, 850, 700, 500, 300, 200, 150, 100, 50, and 25 mbs. Naturally these levels are not analyzed by all stations. Only the best-equipped weather offices would produce these charts. The 500-mb chart will give a fair horizontal indication of the placement of the jet stream, in the absence of a 300- or 200-mb chart. Another basic difference between the surface chart and charts above the 700-mb level

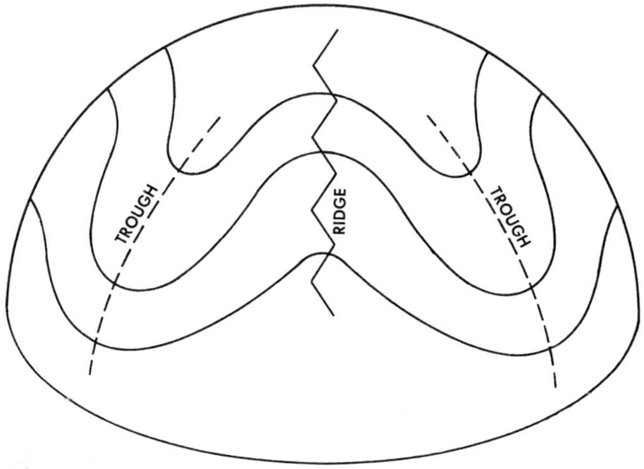

Figure 18-10 Upper-Air Wave Patterns

is that the discontinuities which characterize the surface front are not strong enough to represent them on the analysis.

Where fronts are conspicuous by their absence, the trough and ridge predominate. The latitudinal undulation of the contour lines makes the analysis appear to be a series of troughs and ridges. Closed high and low patterns are less frequent than on the surface map. Those that are seen may be associated with lows and highs on the surface chart. The waves in the contour lines are classified as either *long* or *short waves.* The wavelength of a long wave may vary from 60 to 120 degrees of longitude. The short waves appear to move more rapidly than long waves and are often hard to distinguish since the upper-air data are normally only taken twice a day, at 0000Z and 1200Z.

With the advent of the space age, satellites have proven to be a very valuable adjunct to the other tools of the weatherman. Each successive launch of meteorological satellites brings greater sophistication and effectiveness to the data gathering. The orbiting platform is not a replacement for other tools, or the answer to all problems, but it has made significant contributions, particularly in the field of tropical storm surveillance. The observations by these satellites are also the basis of the *nephanalysis* (Greek, *nephus*—cloud) (horizontal weather depiction or cloud-cover charts). These charts are subjectively drawn by experienced analysts and transmitted by facsimile to interested stations. Experiments with infrared measurement may also have significant results in the measurement of sea surface temperature and cloud cover at night. In addition there is some indication that surface wave and current information may be obtained from the use of satellite-mounted radar. Some Navy ships have the capability of receiving a picture of the weather as the satellite passes overhead. This is very valuable, particularly at sea where data are scarce.

The vertical structure of temperature has been shown to be of the essence of many meteorological processes and gives the meteorologist another important clue to the three-dimensional make-up of the atmosphere. The principles involved were discussed in Chapter 7. The various conditions of lapse rate which promote stability and instability are closely examined on a thermodynamic chart. The information derived is of great assistance in making a complete analysis of meteorological elements. One example is an *arowagram*, which serves much the same purpose in the atmosphere, as does the T-S-σ_t diagram for ocean applications. Figure 18-11 is a sample of an arowagram with a representative temperature trace.

Electronic technology has created the computer and its impact on the weather office has been great. The equations of hydrodynamics have long been known, but the applications of these equations are laborious

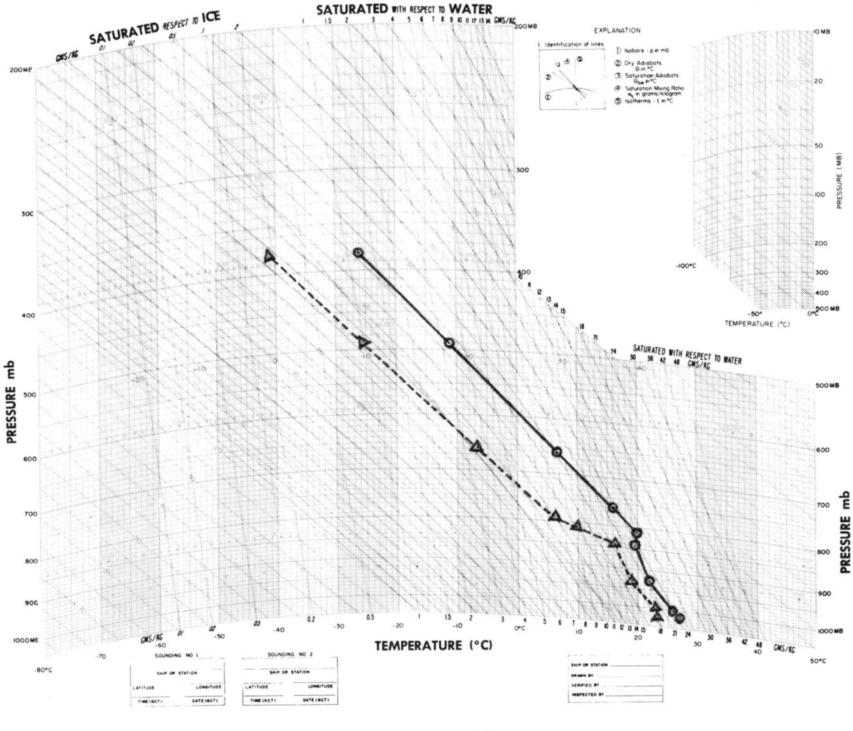

Figure 18-11 Arowagram

Figure 18-12 Plotter Readout of Computer Data

and time-consuming. To utilize these equations in prognosis would require weeks of full-time computations before the single final result would be available. Obviously before this time elapsed, the operations would have been disrupted. The computer, with its fantastic computational speed and ability to reduce the drudgery of routine calculations, has enabled these dynamic equations to be put to good use.

It is estimated that 300 million calculations are required to produce a surface analysis. Computers have been programmed not only to analyze charts but to update their own analyses. Because the synoptic data are so readily adaptable to numerical processes, computerization has been easy. Many products of the Navy's Fleet Numerical Weather Facility, at Monterey, California, are provided on a daily and hourly basis to important weather centrals and even to commercial fishing facilities throughout the world (see Section 1415).

1806 Interpretation and Forecast. The final step in the synoptic process is the formulation of an intelligent forecast, or prognostication (shortened to "prog"). In the case of the professional it requires the mental integration of many factors and indicators. The process is quite similar to that used by a doctor or a stockmarket analyst. All of the symptoms or trends must be checked, evaluated according to their relative importance, and integrated, to formulate a conclusion. Stating it in the most basic way, the forecaster must:

Be alert to situations that may develop weather-producing systems.
Forecast the movement of these and existing systems with respect to a particular location.

In addition to the synoptic picture, the professional forecaster will use many empirical methods, peculiar to a certain location. These have been developed over the years by experienced men at the location in question. For example, a certain combination of temperatures and dew points, coupled with wind direction from a particular sector at Point Conception, California, will almost always produce fog at Santa Barbara, if these conditions prevail at a certain time of the day. Rules such as these come with experience and are handed down to succeeding forecasters.

The amateur forecaster also can make intelligent forecasts with a minimum of assistance. The surface map or chart should be the laboratory where the basic principles discussed in this book are put to practice. The forecast movement of systems should be based primarily on their past performance unless other factors seem to dictate otherwise. Orographic barriers, or the mature stages of an occlusion in the case of wave cyclones, can cause a radical change in rate and direction of a systems movement. The necessary conditions for the formation of radiation fog have been delineated. Any station on the chart showing a very small (one or two degrees) spread between the temperature and the dew point, no low or middle clouds, and light winds would be in an excellent position for a fog occurrence, if the temperature could reasonably be expected to drop another degree or so.

It should be readily apparent that the time of day has a great deal to do with the forecast conditions. In the aforementioned case, if the time were a few hours prior to sunrise, the cooling might occur. However, if the map time indicated the observation was taken just after sunrise, the likelihood of any further cooling would be remote. The ability to mentally convert map time (Greenwich) to local time is a necessary faculty. The Naval Weather Research Facility has published a world chart which shows the hours of daylight and darkness at synoptic times, by season. This chart can be a great aid in visualizing areas of heating and cooling.

The weather sequence associated with frontal activity should be foremost in the mind of the forecaster. A working knowledge of the expected weather can produce an excellent forecast from local observations alone without the necessity for a formal analysis. Cirrus clouds normally appear about 700–1,000 miles ahead of a warm front.

By applying the average rate of motion of a warm front (10–15 knots), the arrival of the front can be predicted. The forecast will begin to verify itself with the appearance of the next type of cloud associated with an impending warm front, the cirrostratus and altostratus. Finally about 250–300 miles (15–30 hours) prior to the arrival of the frontal zone itself, the precipitation will begin. Fog may also shortly precede the arrival of the front. The frontal passage will be accompanied by a veering of the wind from southeast to southwest, a gradual rise in temperature, a drop and then a fairly steady barometer, and gradual clearing. This is the classical sequence of warm frontal passage.

The same process can be applied to a cold frontal passage. Summarized, it is as follows: a fairly sharp temperature fall, a rapid drop and rise in the pressure, veering winds from southwest to northwest, and showery precipitation in a narrow weather band (short time span compared to the warm front). Cumulonimbus and thunderstorm clouds are characteristic of unstable warm air ahead of the cold front.

It should be pointed out that every front does not produce precipitation or even clouds. The definition of a front as the boundary between two air masses allows that the boundary may be quite broad or ill-defined between two nearly similar air masses. In these instances, as mentioned before, the change in the dew point (differing humidity characteristic) may be the most definitive feature of the front.

An ability to forecast the weather can be a skill of particular value to the mariner. The mere planning of ship's work can be closely tied to the weather forecast. Impending weather signs can determine that chipping and painting an area of deck or topside equipment would be a poor bet. Knowledge that the ship will pass through a cold front can advance the scheduling of tasks that will be affected by it. Even the stores officer will be interested in the temperature, for this will have a great effect on the sequence and priority for striking frozen goods below.

The mariner can get a basic surface analysis by radio. It is called the "canned map." It gives the description of the chart and locates and characterizes (i.e. filling, deepening, etc.) pressure centers, fronts, etc., with accompanying isobaric pattern (significant isobars, sufficient to delineate the shape of the system). This is all done over the teletype circuit by identifying the latitude-longitude coordinates sufficiently to locate the features. The chart can easily be plotted in a short time. In addition to the analysis, a "canned prog" is available. The combination of these two charts should give any vessel that is operating independently the ability to determine a pretty good picture of the general weather scheme.

Where is this information available? Who can tell what frequency it will be broadcast on, and at what time? Probably the best friend the mariner has to assist him in coping with the elements is a publication called *Worldwide Marine Weather Broadcasts* published by the Department of Commerce in conjunction with the Department of the Navy. It contains a wealth of information ranging from a basic discussion of weather analysis, to information on ways to identify a forecast of magnetic storm activity or to obtain a warning of sea ice locations. With this compendium a ship at sea need never be without information on the state of the weather, as long as she is able to communicate.

The environment affects a multitude of man's endeavors. The ability to cope with its capriciousness can be a valuable asset. For both the seaman and the airman, weather can be, and often is, an essential ingredient in the accomplishment of his primary task.

Additional Reading

Duthie, W. D., *Notes on the Analysis of Weather Charts*, 3rd ed. (Unpublished), 1964.

Fritz, Sigmund, "Meteorological Satellites in the United States," *American Geophysical Union*, Vol. 44, No. 2, June, 1963.

Huschke, Ralph E., *Glossary of Meteorology*, American Meteorological Society, Boston, Mass., 1959.

O'Connor, James F., *Practical Methods of Weather Analysis and Prognosis*, NAVWEPS 50-2P-502, Government Printing Office, Washington, D. C., 1952.

———, *Aerographers' Mate 1 & C*, Bureau of Naval Personnel, NAVPERS 10362, Government Printing Office, Washington, D.C., 1965.

———, *The Day-Night-Twilight Chart*, NWRF 00-0666-119, Naval Weather Research Facility, June, 1966.

———, *Meteorology for Naval Aviators*, NAVAER 00-80U-24, Government Printing Office, Washington, D.C., 1958.

———, *Radio Weather Aids* (H. O. Pub. 118), Naval Oceanographic Office, Government Printing Office, Washington, D.C., 1963.

CHAPTER NINETEEN

Wind Waves

1901 Introduction. The winds blow, waves grow, and the ship rolls and pitches, so much so that the landlubber and sometimes the salty sailor get seasick. This sequence of events is probably the most evident environmental phenomenon to the apprentice seaman.

In the ocean, waves are everywhere at all times, consequently waves have been studied for many years. During the nineteenth century, wave problems were studied by scientists who very often never even saw the ocean, and many who did threw up their hands at the apparent impossibility of describing the sea surface in mathematical terms. It was said, as a matter of fact, that the only predictable thing about the sea surface was its complete unpredictability.

Nevertheless, during the nineteenth century a number of mathematicians attempted to solve some greatly simplified wave problems. Their findings, although having seemingly little or no relation to the real

Thirty-foot breakers at Oahu
Courtesy: John Titchen, Honolulu, Hawaii

sea surface, have become extremely useful in modern theory. Consequently, some time will be spent initially examining these solutions so that the results can be used later in describing the ocean surface.

1902 Basic Wave Parameters. If a typical water wave is represented by a sine curve, the basic parameters may be defined in much the same manner as was done in Chapter 8 for electromagnetic and sound waves. Thus the *wavelength* (L) is the distance between two corresponding points on the curve having equal phase, the *amplitude* (A) is the distance from the undisturbed surface to the maximum displacement in the vertical direction, and the *wave height* (H), in this case, is simply twice the amplitude. The time it takes for one wavelength to pass a given point in space is called the *period* (T), while the reciprocal of this, the number of waves passing a point per unit time, is the *frequency* f, $f = 1/T$. Since most wind-generated waves of interest have periods from about 5 to 15 seconds, it is more convenient to talk about ocean waves in terms of periods than frequencies.

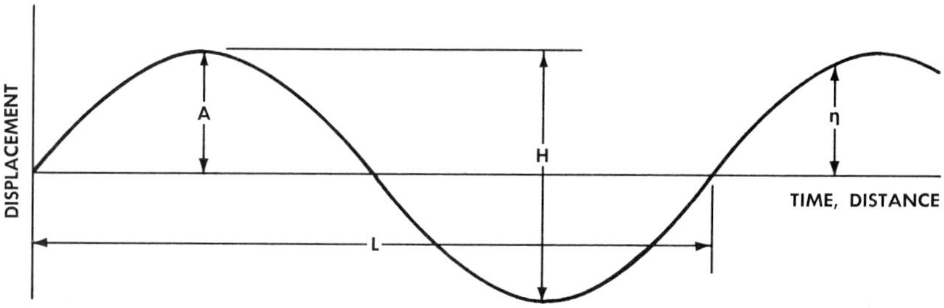

Figure 19-1 Basic Wave Parameters

A very simple and useful relationship is involved with the *wave speed* (c). This is the speed with which the surface profile moves and is related to the wavelength and period by:

$$c = \frac{L}{T} = Lf. \tag{1}$$

This relationship is true for all waves.

Since a progressive wave is moving in space, any attempt to describe the displacement of the sea surface about the equilibrium position must be a function of both distance and time. In general, a sine wave may be represented by the following equation:

$$\eta = A \sin\left(\frac{2\pi x}{L} - \frac{2\pi t}{T}\right) \tag{2}$$

where: η is the instantaneous displacement from the equilibrium position,
x is the horizontal distance measured from an arbitrary origin, and
t is the time measured from an arbitrary reference time.

This equation represents the instantaneous displacement of the sea surface from its equilibrium position in the form of a sinusoidal wave of amplitude A, period T, and wavelength L, which is moving in the direction of the positive x-axis at a speed given by $c = L/T$.

Ocean surface waves may be generated by various forces including wind, tidal forces, seismic disturbances, and atmospheric pressure disturbances. A restoring force that tends to return the sea surface to its equilibrium position, is an essential ingredient for ocean wave production. Without it, no wave will result when the sea surface is distorted by some generating force such as the wind. The restoring forces which are present at the ocean surface are *gravity* and *surface tension*. Since gravity is by far the more important restoring force, in this chapter the discussion will be primarily concerned with wind-generated gravity waves.

1903 The Airy Wave. If a sinusoidal profile is assumed for the wave, and if its amplitude is very small compared to both its wavelength and the water depth, the resulting solution to the wave differential equation is the one proposed by G. B. Airy. The Airy wave predicts a wave speed given by the following expression:

$$c = \sqrt{\frac{gL}{2\pi}\tanh\left(\frac{2\pi d}{L}\right)} \tag{3}$$

where: d is the water depth,
g is the acceleration of gravity (32.2 ft/sec^2), and
tanh is the hyperbolic tangent.

The hyperbolic tangent is a mathematical function similar to the trigonometric tangent function. The basic difference between the tangent and the hyperbolic tangent is that the hyperbolic functions are generated by a hyperbola while the trigonometric functions are generated by a circle. Oftentimes the trigonometric functions are called circular functions for this reason.

When the argument of the hyperbolic tangent is very large, the hyperbolic tangent is approximately equal to one. On the other hand when the argument of the hyperbolic tangent is small, the value of the hyperbolic tangent is approximately equal to the argument itself. Consequently, it appears that two approximations may be made in Airy's wave speed equation, one of these for the case where the ratio of the depth of the water to the wavelength (d/L) is large and the other for the case where the ratio d/L is small.

If the approximation is made that the ratio of d/L is large, the wave speed is given by the following:

$$c\left(\text{for } \frac{d}{L} \text{ large}\right) = \sqrt{\frac{gL}{2\pi}} = c_d. \tag{4}$$

On the other hand, if the approximation is made that d/L is small, the wave speed is given by:

$$c\left(\text{for } \frac{d}{L} \text{ small}\right) = \sqrt{gd} = c_s. \tag{5}$$

If the argument is large, the resulting wave speed describes what is called a "deep water" wave; conversely if the argument is small, the resulting wave speed describes what is called a "shallow water" wave.

In practice, if d/L is greater than 1/2, a "deep water" wave exists, while if d/L is less than 1/20 a "shallow water" wave exists. In between d/L values of 1/2 and 1/20 intermediate waves are said to be present and no approximation is possible. A graphical solution is usually used.

Note very carefully that a "deep water" wave is not necessarily a wave in deep water. The depth involved here is the relative depth since it must be compared with the length of the wave involved. For example,

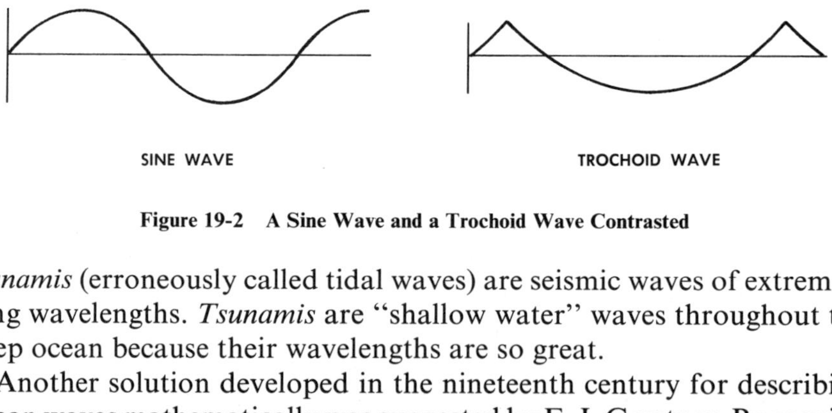

Figure 19-2 A Sine Wave and a Trochoid Wave Contrasted

tsunamis (erroneously called tidal waves) are seismic waves of extremely long wavelengths. *Tsunamis* are "shallow water" waves throughout the deep ocean because their wavelengths are so great.

Another solution developed in the nineteenth century for describing ocean waves mathematically was suggested by F. J. Gerstner. By varying his own basic assumptions slightly, Gerstner ended up with a trochoidal, rather than a sine wave profile. These two curves are compared in the figure. Gerstner's solution is for a "deep water" wave and results in a wave speed identical with that of Airy, i.e.

$$c = \sqrt{\frac{gL}{2\pi}}.$$

1904 Wave Speed. Note that for "deep water" waves, the wave speed is related to the wavelength; long waves travel faster than short waves. For "shallow water" waves, however, all waves travel with the same speed, but this speed is related to the water depth. As shall be seen later, this basic distinction will be of prime import when an attempt is made to forecast wave conditions.

Very often in technical manuals one sees the following expressions for wave speeds and wavelengths in "deep" water:

$$\left.\begin{aligned} c_d \text{ (feet per second)} &= 5.12T \\ c_d \text{ (knots)} &= 3.03T \\ L_d \text{ (feet)} &= 5.12T^2 \end{aligned}\right\} T \text{ in seconds} \qquad \begin{aligned}(6)\\(7)\\(8)\end{aligned}$$

The expression for the wave speed as a function of the period may be easily derived by substituting for the wavelength ($L = cT$) in the initial formula for wave speed. This is done below:

$$c_d = \sqrt{\frac{gL}{2\pi}} = \sqrt{\frac{g}{2\pi} c_d T}$$

$$c_d^2 = \frac{g}{2\pi} c_d T$$

$$c_d = \frac{g}{2\pi} T = 5.12T$$

where: c_d is in feet per second, and
T is in seconds.

The expression for wave speed in knots as a function of the period simply results from the fact that 1.69 ft/sec = 1 knot. That for wavelength as a function of period is deduced very simply also:

$$c_d^2 = \frac{L_d^2}{T^2} = \frac{gL_d}{2\pi}$$

$$L_d = \frac{g}{2\pi} T^2 = 5.12T^2$$

1905 Particle Motion. As a wave passes a given point, the disturbed water particles move in direct response to the wave. A cork bobbing around goes through the same motions as a surface water particle. This is, of course, quite different than the motion of the wave itself since after the wave profile passes on to disturb other water particles the orbital motion ceases and the cork returns to its initial position. But while the wave is causing orbital motion, the type of motion produced is somewhat different for "deep water" waves than it is for "shallow

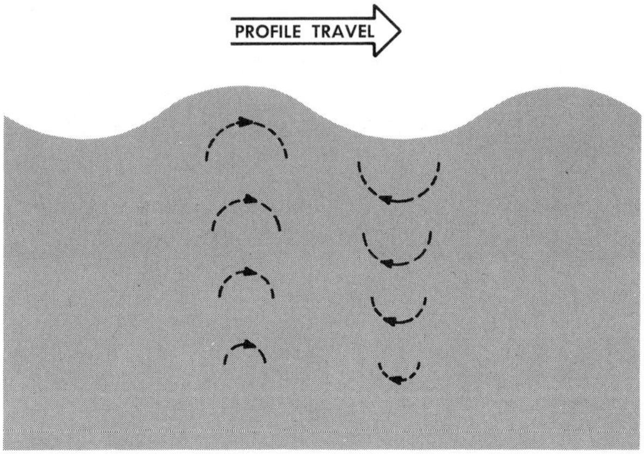

Figure 19-3 "Deep Water" Wave Orbital Motion

water" waves. With a "deep water" wave, the orbits are circular; the size of these circles decreases exponentially with depth. By the time a depth equal to 1/2 the wave length is reached, the size of the orbit is decreased to 1/23 of that at the surface. A common engineering rule of thumb is that in the deep ocean, wind-generated gravity waves are not felt at depths greater than 1/2 the wavelength of the waves. A submarine may very easily escape surface wave effects by submerging a few hundred feet.

In "shallow water," the wave produces orbits which are elliptical rather than circular. The vertical motion decreases linearly to zero at the bottom, whereas the horizontal motion, remains the same from top to bottom. This is in direct contrast with "deep water" waves wherein both the vertical and horizontal orbital motion dies out very rapidly with distance from the surface.

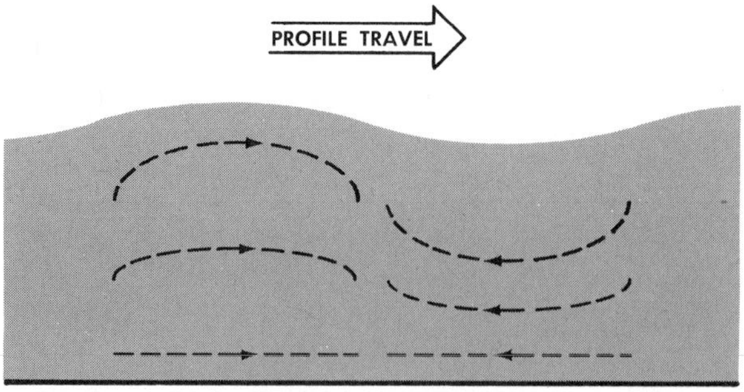

Figure 19-4 "Shallow Water" Wave Orbital Motion

Since the back-and-forth water motion produced by "shallow water" waves is as pronounced at the bottom as it is at the top, this is an important mechanism in sediment transport. These oscillatory currents not only move sand and silt in beach areas, but they also either keep in suspension, or actually put in suspension, a large amount of the sediment moved around in shoal areas.

For all waves, it is usual to say that there is no water transport with the wave; that is, the orbits produced by the individual water particles are closed curves. However, G. G. Stokes, another nineteenth-century in-

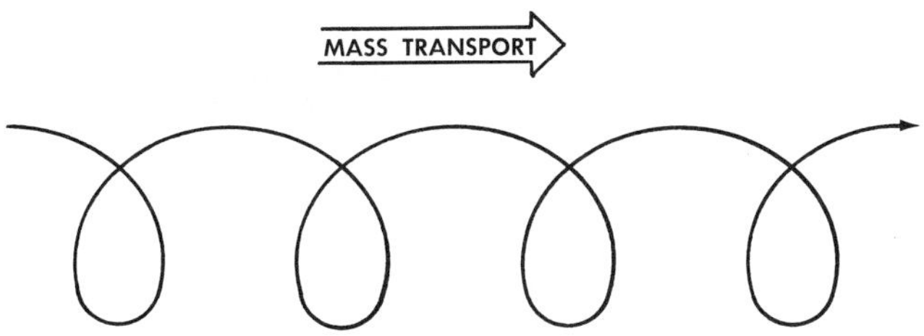

Figure 19-5 Orbital Motion of Stokes' Wave

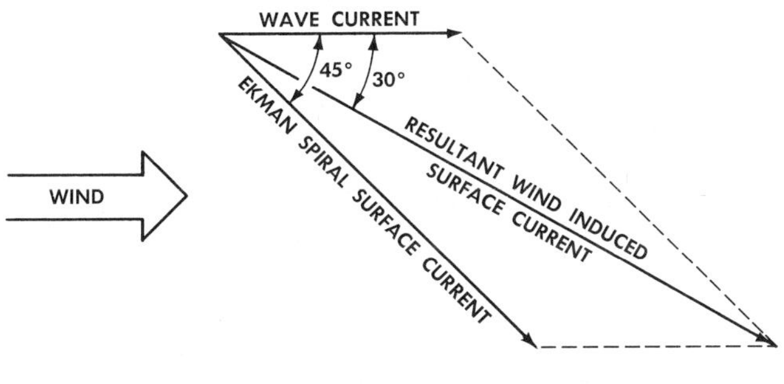

Figure 19-6 Possible Explanation of Wind-Induced Surface Current Not Being at 45° to the Wind

vestigator, suggested a mathematical solution which allowed for a small mass transport of water, or a wave current moving in the direction of the waves. This would be produced as a result of the water particles tracing out open curves in their individual orbits. Since waves generally move in the direction of the wind which generates them, the additional surface current would be in the direction of the wind and must be vectorially added to the surface current produced in conjunction with the Ekman Spiral. This is certainly one possible explanation for the fact that wind-induced surface currents are usually found at about 30° to the wind, but very rarely at 45°.

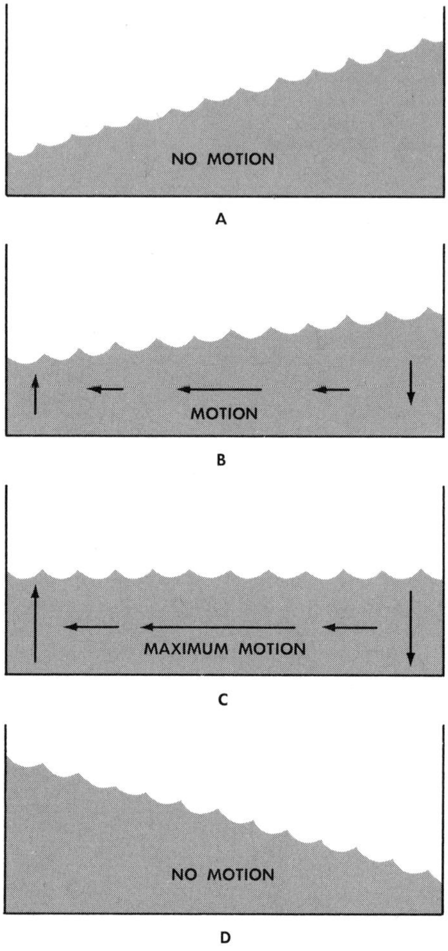

Figure 19-7 Four Stages of a Seiche

1906 Standing Waves. If two progressive waves having the same period and wavelength are caused to move in opposite directions so as to mutually interfere, the result is a standing wave. This may be simply produced by having a wave reflect from a barrier, as is sometimes experienced with wind-generated gravity waves when the waves reflect from a cliff. If this occurs, the resulting standing wave condition is called *clapotis*. With longer waves such as those associated with the tide, the condition of standing waves is a very important one in describing the actual motion experienced. These shall be examined a little more closely in the next chapter, but at this point something of the orbital motion of a standing wave will be considered.

If a tank of water is caused to slosh back and forth, it will set up a standing wave called a *seiche* (pronounced saysh). In the case of a tank, the primary waves set up will be such that the water will produce vertical motion such that there will be a maximum at the two ends of the tank, while in the center there will be none. The reader may try this in a bathtub or a wash basin and observe that at the ends of the tank there can be no horizontal motion since the water cannot penetrate the sides, whereas in the center of the tank there is certainly nothing to hinder horizontal motion.

When the water is piled up on one end to its maximum height, the surface configuration is as given in Figure 19-7A, and at this instant there is no motion in any direction. As the water starts to flow downhill, there is water movement such that at the left-hand end water is starting to pile up and increase the depth of the water (vertical motion up), while it is leaving the right end, decreasing the water depth (vertical motion down). Throughout the rest of the tank there is horizontal motion to the left, with the greatest amount being at the center (Figure 19-7B).

After some time, enough water has moved to produce the condition indicated in Figure 19-7C. A level surface, equivalent to the equilibrium position is present, but maximum velocities exist in all portions of the tank, still directed down on the right, up on the left, and from right to left in the center.

In Figure 19-7D the water has been piled up to its maximum value on the left-hand side; for a brief instant there is no motion of the water anywhere in the tank. The next motion will be toward the right, of course, as the water attempts to regain its equilibrium position. In the course of this movement it should be noted that there never is any horizontal flow at the ends since water cannot flow through the walls of the container. However, these end regions are typified by large vertical flows. Areas of large vertical flows are called *antinodes* or *loops*, and those of no vertical motion are called *nodes*. At the nodal points there is no vertical motion but the largest amount of horizontal motion is experienced. For this elementary tank, a node may be found right at the center.

1907 The Generating Area. With this basic information on wave motion, assume that you are at sea where you can see waves raised by the wind, follow their travel toward the beach, and observe their dissolution as mechanical energy is converted into heat when the waves break. Imagine an area over the ocean surface where the wind is blowing in a particular direction and has been blowing for some time. The energy that the wind puts into the sea surface in the form of surface waves is determined by three parameters: *wind speed*, *wind duration*, and *fetch*. (The fetch is simply the area and significantly the length over which the wind blows in the same direction and at the same speed.)

Empirical studies have determined that for each wind speed there is a maximum amount of energy which can be transferred to the sea surface. Any additional energy is simply dissipated into heat as the waves break. If the maximum amount of energy which can be utilized in the production of waves for a particular wind speed is present, a *fully developed sea* is said to exist. Stated another way, if the wind is putting in more energy than is being dissipated, the waves will grow, but as soon as the dissipation is equal to the input, the waves stop growing and the sea is at its fully developed state. To obtain a fully developed sea (sometimes called a fully arisen sea) minimum values of both fetch and duration are required. Typical values are given in this table:

Table 19-1 Minimum Fetch (Fmin) and Minimum Duration (Dmin) Required for a Fully Developed Sea with Various Wind Speeds (W)

W (Knots)	10	20	30	40	50
F min (Nautical Miles)	10	75	280	710	1,420
D min (Hours)	2.4	10	23	42	69

It is interesting to observe the tremendous increase in both fetch and duration required to produce a fully developed sea as the wind speed is increased. From what has been learned about weather systems, it should be apparent that there are relatively few places on the earth where fully developed seas may be produced for wind speeds over 40 knots. Two examples of areas of extreme fetch are the South China Sea under the influence of monsoon winds, and the southern ocean area around Antarctica under the influence of world-encircling westerlies.

Within the area where the wind is blowing, called the generating area, there is no doubt that waves are being produced. These waves are progressive waves and move out of the generating area in the direction in which the wind is blowing, that is, a west wind will cause waves to move toward the east. Within the generating area the waves are very irregular and confused. Any attempt to find some of the sine waves or trochoidal waves described previously would undoubtedly meet with failure. Not only are the waves non-sinusoidal or non-trochoidal in shape but they appear to have no particular period or common wave height. As a matter of fact they seem to appear and disappear at random. There is no obvious pattern to the sea surface. This randomness within the generating area may be described mathematically by the sum of an infinite number of infinitesimal sine waves, each having a different period, wavelength, and direction. When these sine waves are all added together, they result in the observed sea surface. Note particularly that within the generating area there are a large number of periods present.

1908 Outside the Generating Area. When the waves leave this generating area they are "deep" water ocean waves all mixed together. As seen previously, in "deep" water the long waves travel faster than the short ones. Thus it would be expected that as the waves leave the generating area the situation would become somewhat less confused after a while. Long waves will have outstripped the short waves, causing the wave energy to be spread out over a larger area. This situation where the waves tend to separate due to the fact that different period waves travel at different speeds, is called *dispersion*.

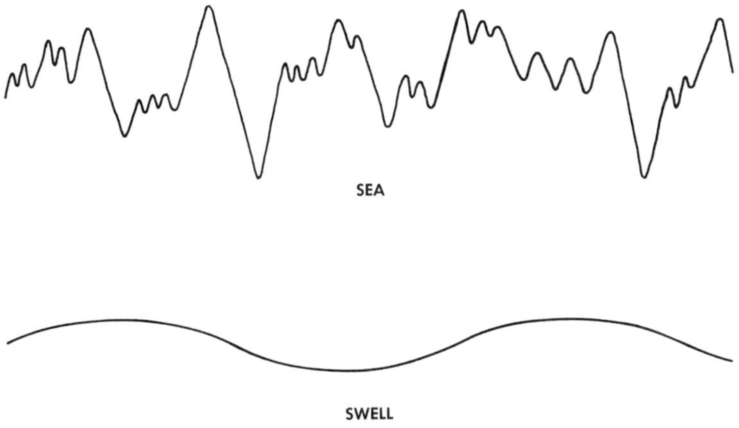

Figure 19-8 Schematic Representation of Typical Sea and Swell Profiles

After the waves have left the generating area and have ceased to be acted upon by the wind, they are then called *swell*. As may be seen from Figure 19-8, swell is characterized by smoothed crests and travels in groups of relatively equal periods. On the other hand, the sharp, peaked waves within the generating area, where all the period components are together, are called *sea*. Due to the fact that swell is spreading over an ever increasing area, the amount of energy per unit surface area contained in these waves is somewhat less than it was in the generating area. This is primarily a result of two phenomena: *dispersion*, as discussed previously, and *angular spreading*. In addition to spreading the energy out along the travel path due to long waves traveling faster than short waves, the energy is also spread out perpendicularly to this path, as if the generating area were initially a kind of point source. This is sketched in Figure 19-9. Since the wave energy is continually being spread over a larger area and energy is related to the wave height, the average wave height tends to decrease with distance from the generating area.

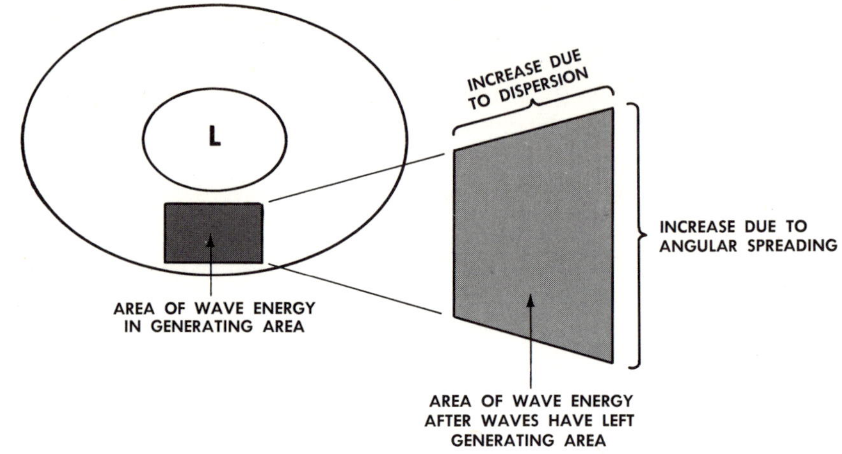

Figure 19-9 The Effects of Angular Spreading and Dispersion on the Wave Energy Per Unit Surface Area

1909 To the Beach. When the waves get into water shallow enough to allow them to take on the characteristics of "shallow water" waves, they will all travel with the same speed. However, as the water depth decreases, this travel speed decreases. In addition to this, the speed of the water particle, as it goes through the orbital motion in a "shallow water" wave, *increases* as the depth of the water decreases. Thus the waves are slowing down as they move toward the beach. At the same time, the speed of the water particles in the orbital motion of the waves is increasing. Eventually a point is reached where the wave becomes unstable. The water particles in the crest simply travel faster than the wave profile, leaving it, causing the wave to break. This breaking apparently occurs when the angle of the crest is less than 120 degrees or when the *wave steepness*, defined as the wave height divided by the wavelength (H/L), becomes greater than 1/7 (see Figure 19-10).

When the waves break, the energy contained by the waves is dissipated and converted into heat. The amount of heat is a significant amount; however, the volume and the heat capacity of the water involved is so large, that any temperature rise of the waters within the surf zone is very small.

There is another phenomenon associated with breaking waves which is quite important. Since water from the breaking wave is carried toward the beach, there must be some way for this water to return to sea. If the waves come in at an angle to the beach, there is a component of movement along the beach called a *longshore* or *littoral current*. This current will move along the beach until it comes to an area where it meets another longshore current coming in the opposite direction. These two will join and move out toward the sea forming a *rip current*. A rip current

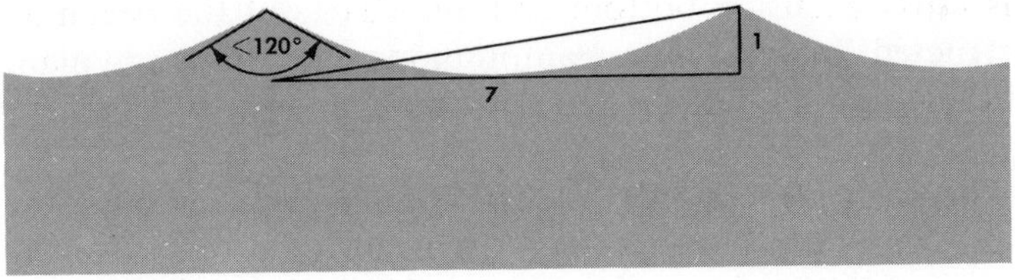

Figure 19-10 Conditions for Breaking

is not necessarily one along the bottom, but is simply a return current directed away from the beach. Rip currents may be dangerous. However, if a swimmer caught in a rip swims across it for a short distance rather than against it, he will find his chances of getting back inshore markedly improved. Rip currents are not to be confused with "undertow," a phenomenon which reputedly drags swimmers down and out to sea but which has never been experimentally verified.

Since the speed at which waves move in shallow water is related to the depth of the water, a pretty good idea of the bottom contours of an area may be obtained by observing the way in which waves are refracted or bent due to variations in speed. Two examples are shown in Figure 19-11. A wave train passing over a bar will result in the waves being bent around the bar, whereas a wave train passing over a submerged valley will find the wave train bent so that the center of the train over the valley travels faster than the regions on either side. Therefore, if the bottom contours are known, the refraction of waves coming from different directions may be calculated.

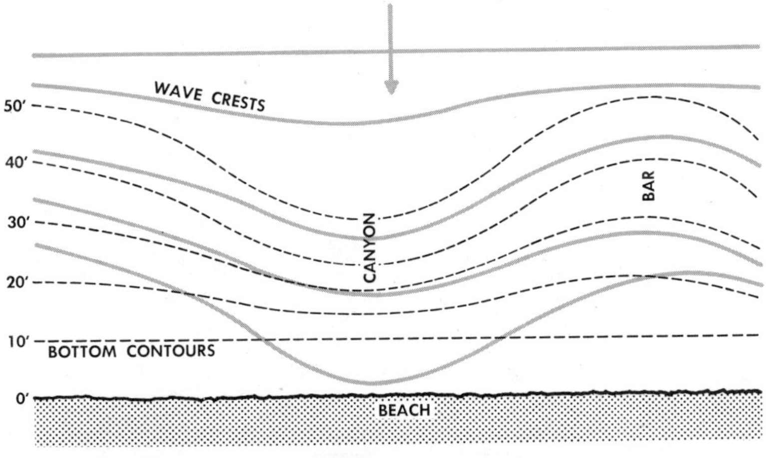

Figure 19-11 The Effect of Bottom Contours on Wave Direction in Shallow Water, as Seen From Above

Conversely, from an analysis of aerial photographs of wave refraction patterns fairly accurate bottom contour charts of the beach area may be constructed. In areas where amphibious operations are planned this is often the only way that accurate bottom contours may be obtained.

1910 The Energy Spectrum. It has been noted previously that wind-generated waves are produced by a transfer of energy from the wind to the sea surface. In forecasting ocean waves, one of the more valuable aids involves the determination of this wave energy which the sea surface contains. For a simple sinusoidal wave the energy contained is proportional to the square of the wave height. If it were possible to go into a generating area and separate the waves into their infinitesimal sinusoidal components, the individual heights could be measured, squared, and then added together. This would give the total energy contained in the sea surface. Obviously, this is impossible. However, a relationship has been found between the three parameters, wind speed, duration, and fetch, and the total energy contained in the sea surface.

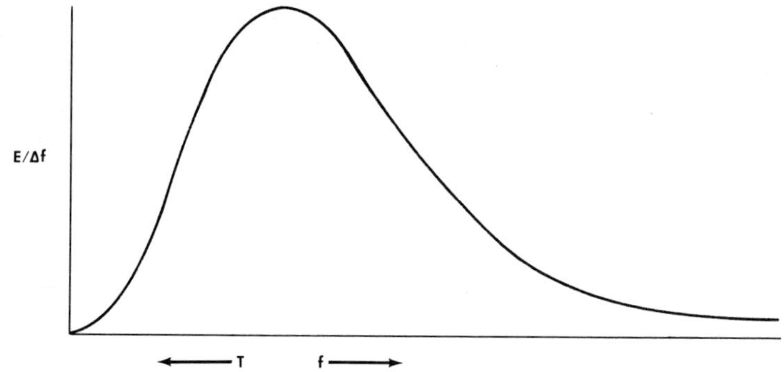

Figure 19-12 A Wave-Energy Spectrum

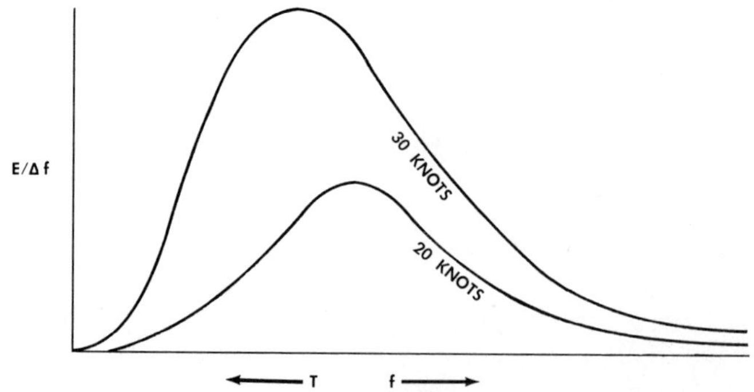

Figure 19-13 Wave-Energy Spectra for Fully Developed Seas Produced by Two Wind Speeds

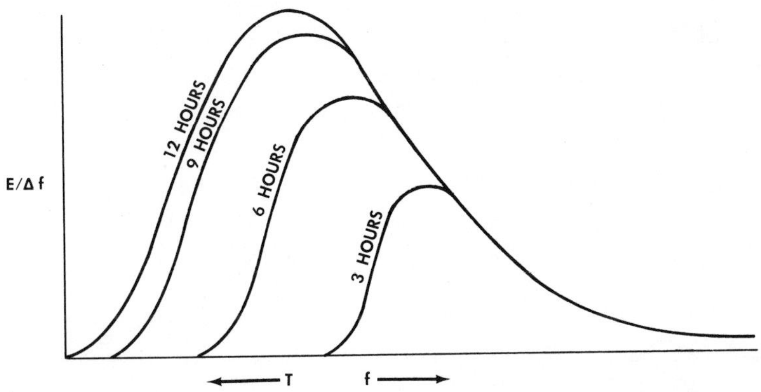

Figure 19-14 The Effect of Wind Duration on the Energy Spectrum

The easiest way of approaching this is to examine the relative amounts of energy contained within different period ranges in the sea surface. Within a generating area where there is a fully developed sea, the relationship between the energy per unit frequency and the frequency is as shown in Figure 19-12. This is called an *energy spectrum*. This particular plot indicates that the energy contained within a band of relatively low-frequency waves is quite high, that contained within a band of lower frequencies drops off, and that contained at the higher frequencies is even less. All frequencies are present but there is more energy within certain frequencies than others.

Now if the wind speed is increased and the fully developed sea which results is examined, not only is more total energy available (high waves), but longer waves (waves of a longer period or a lower frequency) are also present. This is shown in Figure 19-13 where it may be seen that the curve for 30 knots is above, and shifted to the left, of the curve for 20 knots.

As the wind first starts to blow (if the sea is not fully developed), most of the energy is put into short-period waves. The energy spectra for a 20-knot wind blowing for 3 hours, 6 hours, 9 hours, and 12 hours would look as shown in Figure 19-14. Note that the longer the wind blows, the greater the contribution of long-period waves.

The effect of fetch variation is similar, as is shown in Figure 19-15. If a wind of a given speed blows for a long period of time but over a small area, the waves present are relatively short in period (high frequency). If this area is lengthened, the waves present become greater in period so that within protected regions, such as estuaries and lakes, long-period waves never do occur. The fetch always limits their period.

Over the last decade or so energy spectra curves such as shown above have been developed. Therefore if wind speeds, wind durations, and wind fetches are known (these are usually available from weather maps),

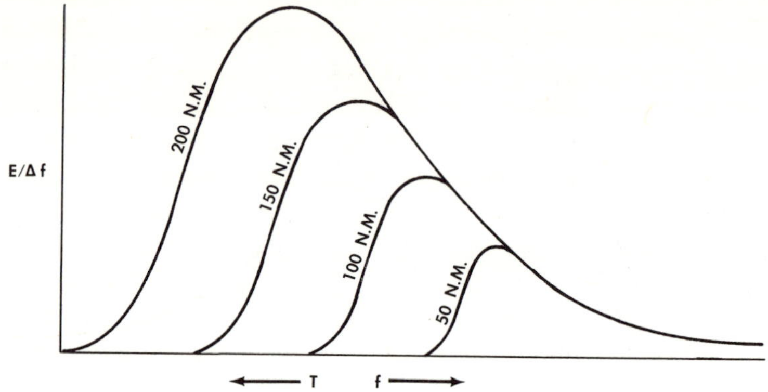

Figure 19-15 The Effect of Fetch on the Energy Spectrum

the various forecasting parameters for the generating area may be determined. The first thing that may obviously be determined from these three parameters is the spectrum curve and the total area under the curve, which represents the total energy. This total wave energy of the sea surface is very imaginatively called E by wave forecasters.

Since the total energy is a function of the wave height, any wave height statistic desired may be calculated. Two typical wave height statistics are average wave height ($\bar{H} = 1.77\sqrt{E}$) and *significant wave height*, the latter the average of the 1/3 highest waves ($\bar{H}_{1/3} = 2.83\sqrt{E}$). Significant wave height is the most frequently reported and forecast, even though it seems that reported wave heights are in inverse proportion to the size of the vessel reporting them. Waves always look bigger from small ships and smaller from the big ships.

In addition to wave heights, the span of wave periods present and the probability of certain periods occurring may be calculated by reference to the energy spectrum indicated before.

Of course, once the waves leave the generating area, modifications result due to the dispersion and angular spreading. Knowing how fast the individual waves travel, from the Airy relations given previously,* the effect of a storm on a distant area at some later time may be predicted. Then, knowing how long it will take waves of a given period to reach a certain beach, and in addition, how much energy is contained within those waves, a pretty fair forecast can be made as to the breakers and surf conditions to be expected.

1911 Wave Measurements. Surface waves may be measured by a number of different devices, the most common of which is the human

* In actuality the group velocity (v_d) is used which is 1/2 c_d. But for this discussion it is sufficient to realize that the Airy relations are of value.

eye. Typically, waves are measured from ships using known values of length and freeboard to determine wave length and height, and a stop-watch for periods. If more accurate data are desired, pressure devices may be installed on the sea bottom or at some predetermined depth to record the pressure effect of surface waves. These have a disadvantage in that the water column tends to filter out the shorter period wave effects especially at depths greater than 1/2 their wave length. Wave staffs may also be used, either attached to the vessel itself, to the bottom, or to some sort of a floating device which is so designed that wave action does not move the staff. Here the interest is in measuring the height of the water about the staff. This can be done either by measuring the electrical capacitance of the staff with respect to the water, or by utilizing a series of electrodes and associated circuitry to determine which electrodes are covered by water and which are not.

Devices containing accelerometers may also be placed on the surface, so that the acceleration in all directions may be continually monitored. This may be integrated once to obtain velocity, or twice to obtain displacement, so that wave parameters may be determined. Submarines may advantageously utilize inverted echo sounders in which sound is sent up toward the surface and bounced from the water-air interface, resulting in a continuous record of the sea-surface profile. This is much the same manner in which a submarine under ice determines its clearance. At the other environmental extreme it appears that there is a distinct possibility that satellite-borne radar may be utilized at some future time to determine wave heights for large areas within very short periods of time. If this is possible, the data input will be greatly increased, and surface winds may be monitored continuously over the entire ocean.

1912 Internal Waves. Waves may exist at any fluid discontinuity, the obvious one being the water-air interface which has been of interest up to this point. However, there is another density interface present in most of the ocean, which has not been considered. This is the thermocline, representing a difference in density between two water masses, which allows for the possibility of waves within the hydrosphere. These waves are called *internal* or *interfacial waves*. Internal waves may be produced by tidal forces, by seismic disturbances, and sometimes apparently by surface vessels if the thermocline is shallow enough.

In the early days of steam power it was evidently possible in some areas to produce internal waves with the ship's propeller in such a manner that all the propulsion energy of the vessel was used to produce waves and none remained for propulsion. This phenomenon has been called "dead water." Nowadays, however, there is usually enough energy available so that "dead water" very rarely occurs.

If a ship can cause internal waves, the amount of energy required to do so is obviously quite small. In fact, the amount required is on the order of magnitude of 1/500 to 1/1,000 of what it takes to produce surface waves of similar height. This is simply due to the difference in density between the underlying and overlying media involved at the two interfaces. Consequently, internal waves may be quite high, 60-foot waves being measured with some regularity. On the average, internal waves have lengths up to 1,000 feet or so, and periods of 15 to 20 minutes are not unusual, so that in general, the wave speeds involved are very low.

Internal waves are a much more common occurrence than had been once thought. Some phenomena associated with internal waves are receiving an increased amount of study. One of these is the production of surface slicks. The Naval Undersea Center (formerly NEL) in San Diego has observed that, in many cases, surface-slick patterns could be correlated with internal waves existing at that time, since the slicks moved right along with the internal waves.

It appears quite possible, then, for a submarine which submerges to avoid the wave action of a surface storm to run into a situation where internal waves cause just as much vertical motion as was experienced at the surface. Of course, in a case like this, a small change in submarine buoyancy may solve the problem, since the disturbance produced by internal waves is often damped out very rapidly in the vertical direction.

1913 Capillary Waves. For the sake of completeness it seems desirable to mention another type of wave which exists at the surface but is usually of small importance. Its major effect is to create a surface roughness which aids in the transfer of energy from atmosphere to sea surface. Capillary waves are generated by the wind, but the restoring force is the surface tension of the water surface. These waves are very small, usually less than a centimeter in length, and their speed is inversely proportional to their wavelength so that the shorter waves travel faster than the longer ones.

Capillaries are the first waves to be formed when the wind blows over a calm water surface, the commonly seen cat's paw being the result of energy transfer from a light breeze to capillary waves. They are the most prevalent of the waves, appearing in rain puddles and bath tubs, and persisting as long as the wind blows.

1914 The Rocking Boat. Surface waves can be both a help and a hindrance, depending on one's point of view. It was probably extremely uncomfortable for the men aboard the USS *Ramapo* in February of

1933 when she encountered a wave 112 feet high (the highest wave ever reliably reported) in the Pacific, but it is from such storm areas that the long period swell so sought by surfers originates. To those aboard an ocean liner, four-foot waves are hardly noticeable; however, a radar operator aboard a smaller vessel might find these waves a real nuisance, effectively masking many targets as a result of sea return.* Yes, to some, rocking the boat brings on a case of *mal de mer;* to others, it is not unlike the soothing motion of a cradle.

* Radar echoes reflected from the sea surface.

Additional Reading

Bascom, W., *Waves and Beaches*, Doubleday & Co., Inc., 1964.

———, "Ocean Waves," *Scientific American*, August, 1959.

Bigelow, H. B., and Edmondson, D. T., *Wind Waves at Sea, Breakers and Surf*, H. O. Pub. No. 602, 1947.

Hill, M. N. (Ed.), *The Sea*, Vol. I, Interscience Publishers, 1962.

Kinsman, B., *Wind Waves*, Prentice-Hall, Inc., 1965.

Neumann, G., and Pierson, W. J., Jr., *Principles of Physical Oceanography*, Prentice-Hall, Inc., 1966.

Pierson, W. J., Jr., Neumann, G., and James, R. W., *Observing and Forecasting Ocean Waves*, H. O. Pub. No. 603, 1955.

Russell, R. C. H., and Macmillan, D. H., *Waves and Tides*, 2nd ed., Hutchinson Publishers, 1954.

Stoker, J. J., *Water Waves*, Interscience Publishers, 1957.

CHAPTER TWENTY

Tides and Other Long Waves

2001 Historical Introduction. The study of tides is one of the few aspects of scientific endeavor which does not have a long history associated with it. No doubt the major reason for this is that the ancient civilizations from whom our cultural heritage is derived were clustered about the essentially tideless Mediterranean Sea. Why that sea is tideless will be seen in the ensuing pages, but for now it is enough to note that the ancient Greeks and Romans were aware of the tides. However, since the tides were so small, they were not considered important enough to be studied.

The first written record of tides and their possible association with the moon is that of Curtius Rufus, the Latin biographer of Alexander the Great, who made a voyage to England in the first century A.D. The tides he encountered there were much too large to be ignored and he managed to relate these with the phases of the moon. When the Romans went to England, they also noticed the connection between lunar phase and the tide, and in addition, were able to identify the difference between spring and neap tides. However, it was not until the thirteenth century that tide tables, or efforts to predict the tides, were attempted, and the seventeenth century arrived before any serious mathematical work was done with the tides. The big breakthrough was Newton's law of gravitation (1687), from which he was able to develop the relatively simple equilibrium tidal theory to account for some of the major tidal fluctuations.

The eighteenth and nineteenth centuries saw additional work of a theoretical nature done by Pierre de Laplace. Later, W. Thomson Kelvin, G. B. Airy, George Darwin, J. W. S. Rayleigh, and Horace Lamb made significant contributions. In the present century very little has been added to basic tidal knowledge, and routine tidal predictions are done essentially by empirical methods rather than by application of theory. An attempt shall be made to see why this is so and to indicate some of the difficulties involved in constructing an adequate, all-inclusive theory.

2002 Equilibrium-Tide Theory. To contrast the empirical findings with the theoretical model, Newton's equilibrium-tide theory will be examined. This will be used as a foundation for the development of a crude, mathematical model of the ocean tides. In this discussion, the rise and fall of water level will be referred to as the *tide*, and the associated horizontal water motion as *tidal current*.

Tides and tidal currents are caused by variation in the gravitational attraction of the sun and the moon on the earth and its water skin. The water is about 4,000 miles closer to (or farther away from) the sun or the moon than is the center of the earth. Consequently, there is a difference between the gravitational force exerted on the water and that exerted on the earth itself, and it is this difference that produces tides and tidal currents.

Notice in Figure 20-1 the gravitational attractive force of the moon on the water directly below it (T_1) is greater than the moon's attraction for the earth itself (T_2). The assumption here is that the earth is completely covered by an ocean of uniform depth. In addition, the water on the opposite side of the earth is attracted toward the moon (T_3) but its attraction is again somewhat less since this water is even farther away. The effect of these three separate forces is to cause the water directly under the moon to move slightly toward the moon and cause the earth to move in the same direction but not quite as much. This produces the water bulge directly under the moon indicated in the sketch. Similarly, the water on the other side of the earth is caused to move toward the moon but not as much as the solid earth. The water is pulled away from the earth on the side toward the moon and the earth is pulled away from the water on the side away from the moon. The net result is a bulge on both sides, directly under the moon and directly opposite.

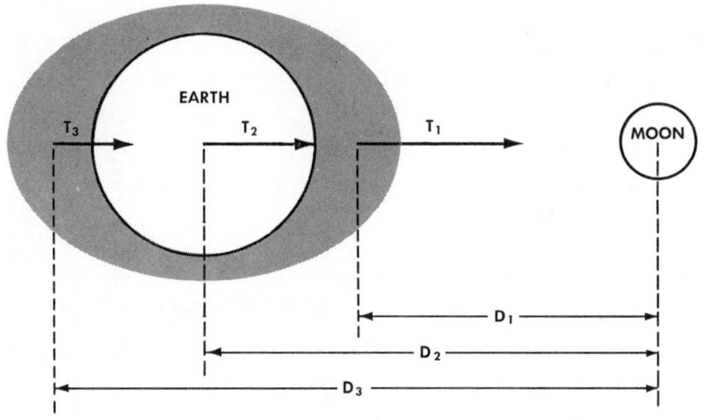

Figure 20-1 The Double Bulge Produced by the Moon

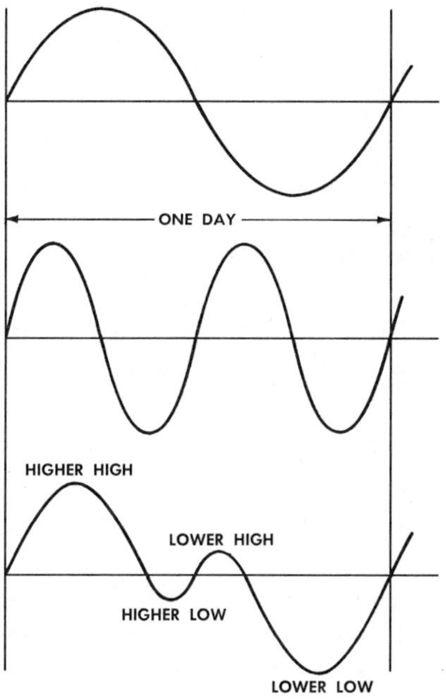

Figure 20-2 Diurnal, Semidiurnal, and Mixed Tides

In other words, with this simple equilibrium tide model there is *a high tide on both sides of the earth.* The relatively simple mathematics associated with determining the order of magnitude of these forces shows that the force actually involved in moving the water to produce this bulge is proportional to the mass of the heavenly body producing the force and inversely proportional to the cube of the distance of the heavenly body. This force is called the *tractive force.**

* The tractive force is the tangential component of the difference between forces T_1 and T_2.

Thus, even though the sun is more massive than the moon, it is so much father away from the earth that the tidal effect of the sun is only about 46 percent of that of the moon. All heavenly bodies have some tidal influence on the earth. However, they are either so small or so far away that their tidal contribution may be neglected. The sun and the moon are the only two heavenly bodies which are large enough and close enough to have any noticeable tidal effect on the waters of the earth.

As the earth rotates about its axis the local water level changes periodically. The moon will be used in a simplified picture to observe this time variation, but certainly the sun's effect is essentially the same. Note that as the earth spins on its axis, the bulge underneath the moon remains fixed, so that once every lunar day there are two high tides and two low tides at each particular spot on the earth's surface. This twice-daily high and low tide occurrence is called a *semidiurnal period.* In addition to the semidiurnal period there are a *diurnal period* occurring once a day and other greater periods in the tide-producing forces, which are all grouped together as *long periods.* A *mixed tide* is said to exist when both diurnal and semidiurnal components are obviously present, which implies unequal highs and unequal lows (see Figure 20-2).

2003 Tidal-Range Variation. If the earth, the sun, and the moon are all in a straight line,* as occurs whenever there is a full or a new moon, the latter two bodies tend to act together producing *tidal ranges*† somewhat greater than normal, called *spring* tides. On the other hand, when the moon, the earth, and the sun are at right angles to each other (a situation called *quadrature*), the bulge produced is less than average in size, since the two are not working together; and a *neap* tide is said to exist. Spring and neap tides occur once every fourteen days, which is one-half the period of revolution of the moon about the earth. This period of lunar revolution is one of the long-period tidal forces which must be considered in any tidal theory.

If the moon is now allowed to change its declination‡ so that a line drawn from the moon to the earth is no longer perpendicular to the earth's axis, the tidal variation is somewhat different. As the earth rotates, the bulge will remain at a fixed latitude, but with the passage of 1/2 lunar day, an observer at this latitude would note a different height when the moon is on the opposite side of the earth. If a section is cut through the earth at a particular latitude, the bulges are seen to be asymmetrical. Whenever the declination is different from zero, the bulge is greater on one side than on the other resulting in an unequal tide.

* This is called syzygy.
† By *tidal range* is meant the difference between high and low tides.
‡ The angle in space made between a heavenly body and the extended terrestrial equator.

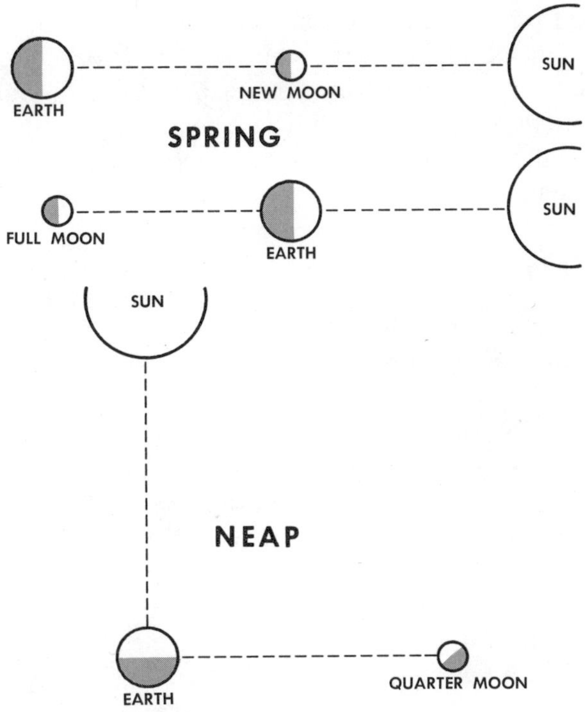

Figure 20-3 Spring and Neap Tides

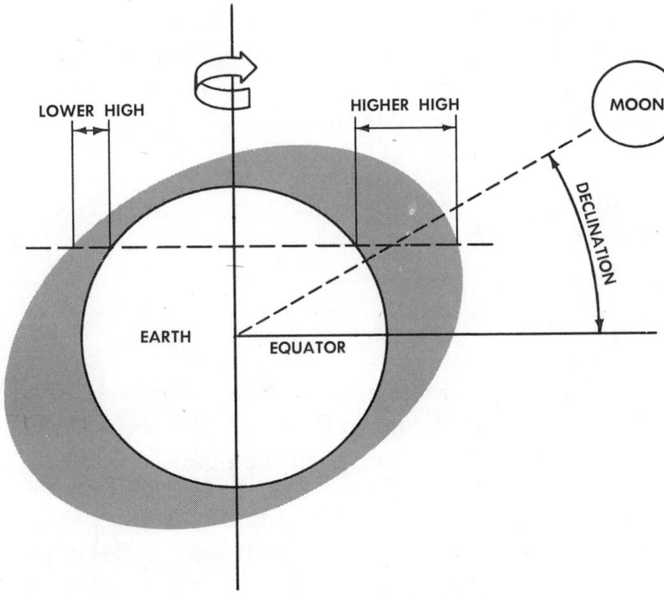

Figure 20-4 Diurnal Inequality

For a period of one lunar day and a declination not equal to zero, there is one higher high tide and one lower high tide, along with one lower low tide, and one higher low tide, rather than two high tides of the same height and two low tides of the same height. This diurnal inequality is more representative of earth tides, since zero declination of the moon occurs infrequently.

2004 Tide Prediction. The forces which cause the tides are therefore periodic and are produced by the motions of the sun and the moon as the earth rotates about its axis and revolves about the sun. A few of the recurring motions that produce tides are:

The moon's revolution about the earth every 27.3216 days.
The earth's revolution about the sun every 365.2422 days.
The moon's oscillation north and south of the ecliptic* as the moon's declination changes every 27.2122 days.
The oscillation in lunar distance as the moon moves farthest away from and closest to the earth every 27.5546 days.
The earth's rotation about its axis with respect to the sun every 24 hours and with respect to the moon every 24 hours and 50 minutes.

These are just a few of the many periods which the sun and the moon go through in the production of tides. One method of predicting tides is to consider all of these periods. Measurements for a particular area of interest are made over a long period of time, and are then analyzed for the presence of astronomical periods such as indicated above. Each of these astronomical periods, called *harmonic constituents*, will have a particular amplitude and a particular phase, which may be expressed as follows:

$$\text{Tidal Height Produced by Period } T_1 = A_1 \cos\left(\frac{2\pi t}{T_1} + \varepsilon_1\right)$$

where: A_1 is the amplitude
T_1 is the period, and
ε_1 is the phase.

Once the amplitude and phase are determined for each of the harmonic constituents present, the actual tide is simply the sum of all these harmonic constituents which have been included.

$$\text{Predicted Tidal Height} = \sum_{i=1}^{n} A_i \cos\left(\frac{2\pi t}{T_i} + \varepsilon_i\right) \qquad (1)$$

* The plane made by the earth's orbit in space.

In practice, somewhere on the order of 40 of these harmonic constituents are usually utilized in a typical tidal prediction.

It must be emphasized that the only way to predict tides is to make measurements for a selected area, and extrapolate these measurements into future times or adjacent areas. Tides cannot be predicted unless data are available to work with.

There are a number of reasons why simple theory does not adequately describe tidal fluctuations. In the first place, the earth is not completely covered with water, nor is the water of constant depth, both of which Newton assumed in developing his equilibrium tide. Secondly, basin size and shape have a large effect on the tide, along with unique coastal configurations. Thirdly, the effect of a rotating earth in producing coriolis deflection is a major cause of discrepancy from the theoretically expected tides. The latter two causes will be briefly examined.

If an enclosed body of water is caused to oscillate, it will be found that a particular period of oscillation is preferred. This preferred period is a function of the dimensions of the body and is called the resonant period. Under conditions of resonance, standing waves are set up and the maximum possible water motions are produced. Since resonance requires the smallest energy input, when a basin is excited by a number of periodic forces, it will tend to respond only to the force having a period identical to its resonant period.

For an enclosed basin such as an ocean or a lake, the type of standing wave oscillation discussed in section 1906 is seen to be associated with a wavelength equal to twice the size of the basin.

$$L = 2l_c$$

Since this is a "shallow water" wave

$$c = \frac{L}{T_c} = \frac{2l_c}{T_c} \quad \text{and} \quad c = \sqrt{gd}$$

so that

$$\sqrt{gd} = \frac{2l_c}{T_c}$$

and the resonant period of an enclosed basin (T_c) is given by:

$$T_c = \frac{2l_c}{\sqrt{gd}} \tag{2}$$

where: l_c is the length of a resonant basin,
d is the average depth, and
g is the acceleration of gravity.

If the basin is in the shape of a gulf, i.e. open at one end, the standing wave is seen to have a wavelength four times the basin size,

$$L = 4l_0$$

so that

$$\sqrt{gd} = \frac{L}{T_0} = \frac{4l_0}{T_0}$$

and the resonant period of an open basin (T_0) is

$$T_0 = \frac{4l_0}{\sqrt{gd}} \quad (3)$$

where the parameters are the same as above. It may be seen that the resonant period of a gulf of a particular length is twice that of a closed sea of the same size.

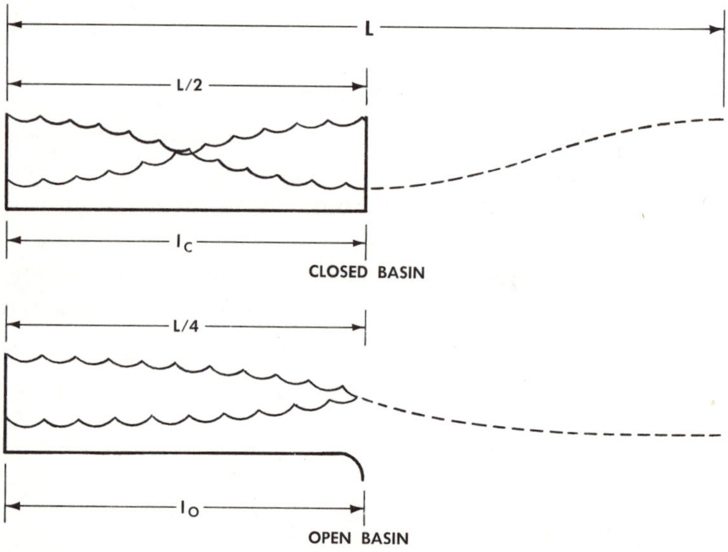

Figure 20-5 Seiche in an Enclosed Basin and in an Open Basin

The important thing to realize about resonance in both gulfs and basins is that if a number of periodic forces are acting together, such as is found with tidal forces, a body of water will respond most actively to the one which is close to its resonant period. Consequently, certain areas of the world have predominantly diurnal tides and other areas have predominantly semidiurnal tides, simply because they respond best to the forces having these periods.

Since tides are periodic, they may be considered as waves, and with the simplest model possible, that of Newton's equilibrium tide, these waves have a wavelength of about 12,500 miles at the equator. These are very, very long waves and even in the deepest part of the ocean they would be "shallow" water waves since their wave length is so much greater than the water depth. The currents associated with tides therefore are simply the orbital motions of the tide wave, and as with all other "shallow" water waves, they extend undiminished to the bottom.

2005 Amphidromic Systems. Since tide waves are so long, the motion associated with them is affected by the rotation of the earth, and coriolis force is an important determining factor in the resulting motion. A particularly important case of the effect of coriolis force results when a standing wave is present in an oceanic basin. Many of the observed effects of the tides can be accounted for by the effect of coriolis force on a standing tide wave. The back-and-forth motion of a standing wave in conjunction with the perpendicular coriolis force pro-

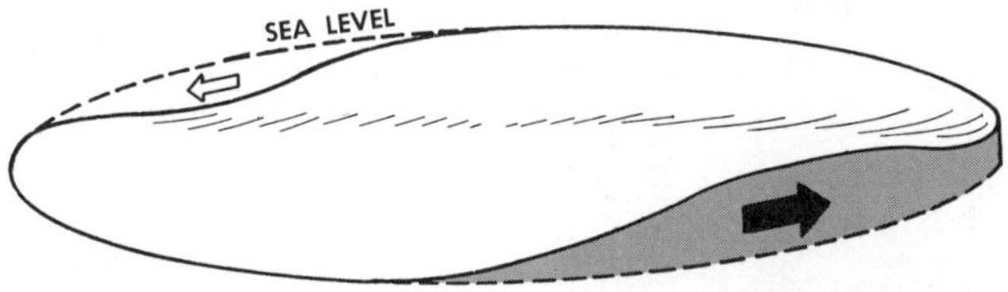

Figure 20-6 A Schematic Representation of an Amphidromic System, Showing the Rotation of High Water about a Central Amphidromic Point

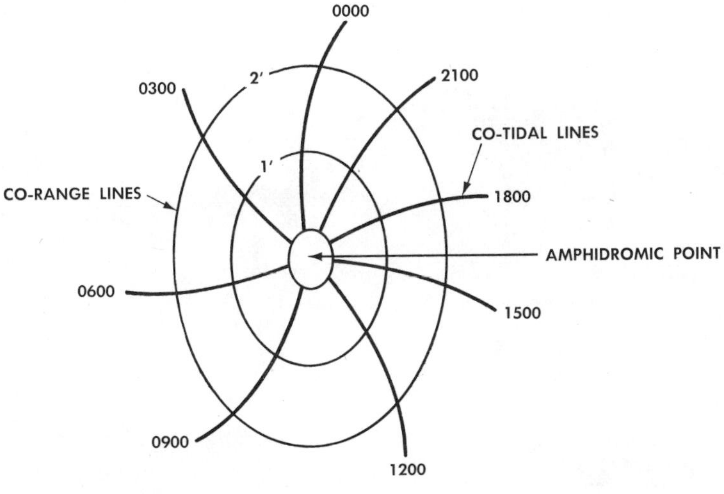

Figure 20-7 Co-Range and Co-Tidal Lines within an Amphidromic System

duces a mound of water which, in the Northern Hemisphere, rotates counterclockwise and, in the Southern Hemisphere, rotates clockwise. Instead of having a simple sloshing back and forth as is found in a normal standing wave, the coriolis effect deflects the motions in such a manner that the antinodes rotate about the central point. Thus instead of having a nodal line as in a normal standing wave, a nodal point is formed. This nodal point is called an *amphidromic point* (Greek: *amphi*—both and *dromos*—running) and the system of these rotating currents is called an *amphidromic system*. One such system is sketched in Figure 20-6, and an attempt is made in the figure to show water heights and water currents which exist in an idealized amphidromic system.

The lines radiating out from the amphidromic point are called *co-tidal lines*. Along any co-tidal line equal phase in the tide exists, so that high tide occurs simultaneously along a co-tidal line. The concentric curves drawn about the amphidromic point, are called *co-range lines*, and anywhere along these lines there will be an equal range in tide. Note that the range in tide increases with increasing distance from the amphidromic point, while at the amphidromic point it vanishes.

2006 Tidal Currents. In some harbor areas such as Liverpool, London, and Anchorage, the tidal range itself is the predominant feature, since it is so great. However, in other areas, tidal currents are the more important; values of ten knots in Seymour Narrows, Alaska, and four knots in the Golden Gate entrance to San Francisco Bay are not unusual. It therefore becomes desirable for the mariner to be familiar with these currents to navigate within a confined area. Since tidal cur-

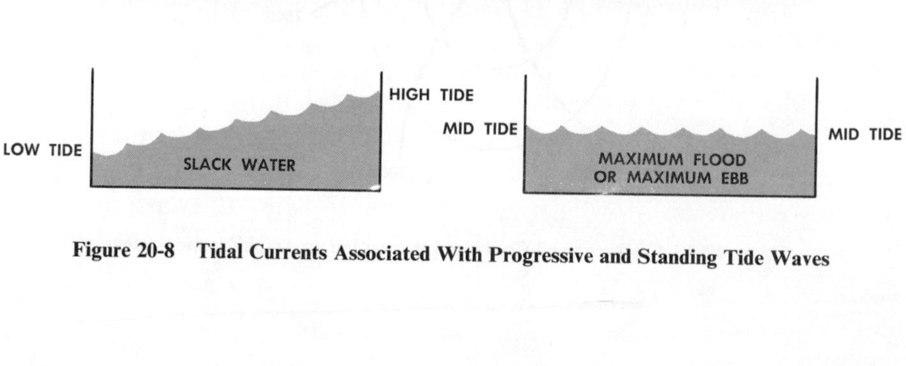

Figure 20-8 Tidal Currents Associated With Progressive and Standing Tide Waves

rents are simply the orbital motions of tide waves, the phase relationship between tidal current and height stages for both progressive and standing waves are of some interest. A progressive wave produces orbital motion in the crest in the same direction as the wave profile motion, and in the opposite direction in the trough. For a progressive wave, high tide will therefore be associated with maximum flood current, and low tide with maximum ebb current.

In a standing wave, however, the situation is somewhat different. As discussed before, the regions of maximum vertical motion are regions of no horizontal current, and regions of maximum current are regions of no vertical motion. Thus, in a standing wave, high and low water will both be associated with slack water, and the maximum ebb and flood will both occur at mid-tide.

In any real situation the tide waves existing are probably going to be some combination of progressive and standing waves, so that there will be some lag between the occurrence of flood and ebb currents and high and low water that must be determined empirically.

2007 Local Anomalies. With the previous material in this chapter as a background, it is pertinent to now look at some of the real tides in the world and attempt to explain their existence. In some areas of the world, very large tidal ranges exist, while in others very small ranges are experienced. Why is this? The place having the greatest tidal range in the world is the Bassin des Mines in the Bay of Fundy in Canada. It has a mean spring range of 44.6 feet. Frobisher Bay, also in Canada, has about the same mean spring range. The Severn River in England has a mean spring range of about 43 feet, and there are a number of others of this nature having mean spring ranges over 35 feet. In all cases, however, these tides occur at the end of a gulf which is resonant to a tidal period. Thus the water sloshes back and forth in the gulf in such a manner that extremely high tidal ranges are recorded.

At the other extreme are regions of the world having very small tidal ranges, an example of which is the previously mentioned Mediterranean Sea. There are probably two major reasons for this small range. First, its dimensions are all wrong to be resonant to any astronomical period, and secondly, the size of the connection between the Mediterranean and the Atlantic Ocean is much too small to allow the passage of a wave into and out of the Mediterranean twice a day, or even once a day. Consequently, the tidal range in the Mediterranean Sea is typically only about 10 to 15 centimeters.

Of the same order of magnitude are the tides at the Solomon Islands in the Pacific, which have spring ranges of about 12 centimeters, and at Puerto Rico in the Atlantic, with spring tide ranges of only three centimeters. In both of these latter cases the islands happen to be located quite near amphidromic points, so that at these nodal points there is very little vertical motion of the water.

Due to the fact that the Atlantic Ocean and the Pacific Ocean are much different in size, the Atlantic being much smaller than the Pacific, the Pacific tends to respond more to diurnal tides and the Atlantic to semidiurnal tides. The result of this is that the west coast of the United States has predominantly mixed tides while the east coast of the United States has predominantly semidiurnal tides.

Thus, even though the tidal forces are known, the tides cannot be accurately predicted without accounting for basin configurations, continental placement, the rotation of the earth, and the individual resonant characteristics of harbor and estuarine areas. It is therefore necessary, as indicated previously, to acquire local data before any tidal predictions can be made.

2008 Meteorological Tides. Other causes of water-level variation and water movement include atmospheric disturbances, generally classified under the heading of meteorological tides, and seismic disturbances. In some areas of the world, local weather conditions are more prevalent in determining water level than tidal effects. This is often true along the east coast of the United States, for example. However, in other regions this is not the case at all. In the Pacific, in the main, tidal heights are determined predominantly by tide forces.

One of the more common of the meteorological tidal effects is the seiche. A seiche consists of standing waves (produced typically by atmospheric-pressure variations or by the passage of a storm system) which are established in closed or semi-closed basins. As each basin has its own resonant period, the standing waves produced within the

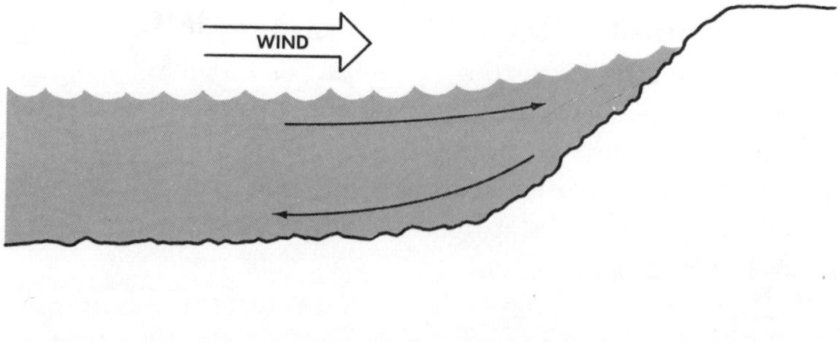

Figure 20-9 The Effect of Bottom Roughness on Wind-Generated Tides

basin have a period depending upon the basin's dimensions. When local conditions suggest the possibility of a seiche, care must be taken to assure proper line-tending by moored ships.

Perhaps the most predominant of the meteorological tides are those produced by the wind. They are especially noticeable in water too shallow to establish the Ekman Spiral effect. In this case the water moves in the same direction as does the wind. Within a long narrow estuary, for example, the water will pile up at the head, if the wind is blowing up the estuary, and will cause exceptionally low tides if the wind is blowing out of the estuary. The amount of water moved by the wind is related to the wind speed and the bottom roughness—the less smooth the bottom the more difficult the return path for the water. A rough bottom facilitates the change of water level by the wind.

2009 Seismic Sea Waves. Another group of phenomena are surges which may be produced by seismic disturbances. These changes in sea level can be catastrophic since occasionally a seismic disturbance will produce a long-period wave train called a *tsunami*. These are erroneously called tidal waves, but having nothing to do with actual tides, the Japanese term has been chosen in order to alleviate the confusion. Unfortunately, *tsunami* means tidal wave in Japanese.

Tsunamis are very long waves, having periods on the order of 15 or 20 minutes and consequently are "shallow" water waves in the deep ocean. They travel at a speed determined by the depth of the water. Speeds of 500 knots are not uncommon. In the deep ocean they rarely exceed a foot or so in height but in isolated cases they will markedly increase in size in coastal areas, probably due to resonant conditions for particular frequencies present in the tsunami. Under these conditions, tsunamis have been known to cause great damage and tragic loss of life. At the present time the Environmental Sciences Service Administration of the Department of Commerce is attempting to perfect a tsunami warning system within the Pacific area. This system has been in operation since 1946. It has been quite successful in warning coastal inhabitants of tsunami approach, but has not been able to differentiate between destructive and nondestructive tsunamis up to this time. Thus the major drawback in the warning system is that everyone is alerted for all seismic waves, whereas very few of them are actually dangerous.

2010 Summary. If an accurate analysis of water height and water current at a particular time and location is to be made, a number of phenomena which affect sea conditions must be considered. In the preceding chapters, you have read detailed discussions of:

Tides
Prevailing currents
Winds
Rapid changes in atmospheric pressure
Motion of weather systems
Seismic disturbances

When all these factors are considered, it should be possible to determine water levels and water currents with a fair degree of precision.

The sea and the atmosphere are extremely complex individually; of course, when interaction occurs, the complexity is compounded. A great deal of progress has been made in the understanding of the marine environment; a great deal is yet to be made. At the present time, however, there is enough pertinent knowledge available to provide the mariner and aviator with significant aid. In addition, it is probably possible for fishing fleets and navies to increase their efficiencies an order of magnitude by intelligent and imaginative use of available environmental data. There will be improvements in the future, but there is no need to wait.

Additional Reading

Bernstein, J., "Tsunamis," *Scientific American*, August, 1954.

Clancy, E. P., *The Tides*, Anchor Books, Doubleday & Company, Inc., 1969.

Darwin, G. H., *The Tides and Kindred Phenomena in the Solar System*, 1898. (Reprinted 1962 by W. H. Freeman & Co.)

Defant, A., *Ebb and Flow*, University of Michigan Press, 1958.

———, *Physical Oceanography*, Vol. II, Pergamon Press, 1961.

Doodson, A. T., *Oceanic Tides, Advances in Geophysics*, Academic Press, 1958.

———, and Warburg, H. D., *Admiralty Manual of Tides*, Her Majesty's Stationery Office, 1941.

Dronkers, J. J., *Tidal Computations*, North-Holland Publishing Co., 1964.

Hill, M. N. (Ed.) *The Sea*, Vol. I, Interscience Publishers, 1962.

Macmillan, D. H., *Tides*, American Elsevier Publishing Company Inc., 1966.

Proudman, J., *Dynamical Oceanography*, Dover Publications, 1952.

Tricker, R. A. R., *Bores, Breakers, Waves, and Wakes*, American Elsevier Publishing Co., Inc., 1964.

FOOD
Harbor Hut

CHAPTER TWENTY-ONE

Epilogue

2101 Understanding the Environment. In the preceding chapters the reader has been introduced to the characteristics and important processes of both the hydrosphere and the atmosphere. Interrelationships between the sea and the air have been emphasized, the transfer of energy and water across the naviface being given special attention because that transfer controls the weather and the climate in both fluids. The terms *weather* and *climate*, as applied to the ocean, refer to the prevailing temperature and salinity conditions and their changes, just as in the atmosphere they refer to temperature and water-vapor values and their changes.

In addition to emphasizing the extremely close coupling between the hydrosphere and the atmosphere, every effort has been made to show that all the various systems within the hydrosphere itself are closely associated. The biological, geological, chemical, and physical systems are all so closely tied together that it is impossible to describe completely any one of them without including some elements of the other three. Oceanography is, as has often been said, an interdisciplinary science.

The primary purpose of this book is to serve in some way those who would place themselves at the mercy of the elements: mariners, sub-

The environment is bound to be changed by the encroachment of people
Courtesy: Baron Wolman

mariners, aviators, and all who work or travel in the marine environment. It is hoped that these people will be inspired to learn how best to use the environment to their own advantage. Communication, identification, and orientation within the environment are affected very strongly by the environment. Therefore, if one is to use one's surroundings or to know what to expect from them one must understand them.

2102 Changing the Environment. It is hoped that the previous chapters have made the reader aware of the fact that neither the sea nor the air is infinite in extent. Since man can change the environment, and he is changing it, any change made should be to his advantage; that is, society must learn how to *manage* the environment.

Both the hydrosphere and the atmosphere are accepting large amounts of waste products at the present time, and all indications point to the fact that they will be required to accept larger amounts in the future. This situation raises a number of problems, some of which will be examined with a view to solidifying various concepts covered in the previous chapters and to indicating how this material can be put to very practical use.

2103 Personal Waste Products. Probably most of the waste products discharged into the hydrosphere and the atmosphere are those that result from personal actions. Among such products are domestic sewage, gaseous discharge from automobiles, fumes from home heating and other individual burning, laundry detergents, and the debris from recreational activities. The greatest single personal pollutant of the atmosphere today results from the use of automobiles, while the largest single pollutant of the hydrosphere results from domestic sewage.

Every gallon of gasoline burned by an automobile (or by any other device) vents about 6 pounds of additional gaseous material into the atmosphere. This gaseous material contains sulphur, nitrogen, water vapor, carbon dioxide, carbon monoxide, various hydrocarbons, and lead, as well as small particulate matter and, of course, heat. Some of these chemicals, especially sulphur, nitrogen, and hydrocarbons, form, usually in the presence of sunlight, active compounds that are damaging to crops and structural materials, and serve to irritate, if not destroy, human tissue. These compounds are primarily responsible for the harmful effects of smog. There is some feeling, however, that the most dangerous, long-term, automobile-derived pollutants are carbon dioxide, carbon monoxide, and water vapor. These three compounds are relatively transparent to visible light and relatively opaque to infrared radiation, so that they tend to increase the greenhouse effect of the atmosphere.

There is a great deal of concern that if fossil fuels continue to be burned at the present rate, our weather and climate patterns will be markedly changed. This is especially disturbing because modern trends toward eliminating automobile pollution by decreasing sulphur, nitrogen, and hydrocarbon emissions tend to decrease engine efficiency and to increase the amount of carbon dioxide and carbon monoxide put into the atmosphere. So, actually, pollution eliminators may serve to eliminate not only the pollutants but the people they are designed to protect.

Domestic sewage dumped into a body of water adds to that water both large quantities of organic nutrients and unwanted bacteria and turbidity. The latter, however, are usually controlled by a primary treatment process, so that their effects are generally outweighed by those of the nutrients. A large input of nutrients often produces basic changes in the ecological community of the water concerned. For example, one plant type will thrive on the nutrients and grow in such profusion that it chokes out all other types, including those on which the animal inhabitants of the water lived. Unable to survive on the "weeds," those animal forms give way to forms that are capable of surviving under such conditions, or animal life in that body of water ceases to exist.

Sometimes the plants grow so rapidly that they completely cover the surface of the body of water, forcing other plants to the bottom of the layer where, excluded from sunlight, they die. These dying plants decay at such a rapid rate that all the oxygen produced in the water by the surface plants is used for biological decay and none is left for the animals. This situation, caused by an excess of nutrients usually supplied by domestic sewage, is called *eutrophication* and it exists at the present time in many lakes and rivers.

Another source of nutrients are the phosphates contained in home laundry detergents, but their contribution to the environment is generally considered insignificant compared with that of domestic sewage. In fact, some scientists believe the nutrient supply from domestic sewage is so great that even if all the phosphates in home laundry detergents were eliminated there would be no noticeable difference in the environment.

Other personal forms of waste disposal include atmospheric pollution caused by home heating and open burning of leaves and charcoal. The effects of all burning are similar to those produced by an automobile engine, so the previous comments are applicable. Pollution resulting from recreation takes diverse forms, from increased silt loads of streams caused by boating and swimming activities to dead fish thrown back by fishermen who have caught either too many or the wrong kind.

2104 Industrial Waste Products. Industrial firms also discharge into the environment many noxious waste products. Just as does the burning of any fossil fuel, industrial smokestacks may discharge into the atmosphere sulphur and nitrogen compounds that can be damaging both to people and to construction materials.

Many substances are added to the marine environment, too. Some of them, such as those from the food-processing industry, for example, are organic. The amount of organic waste discharged by all industries in the United States at this time is almost as much as that resulting from domestic sewage.

Industrial wastes may also be noxious materials such as acids discarded in various chemical industries and metal refining; heavy metals such as cadmium, mercury, and lead; silt from dredging and construction-site runoff; oil such as occurs when a pumping station malfunctions or a tanker has a collision; or drainage from abandoned mines. All of these are potentially dangerous and cause serious problems if not sufficiently diluted.

Another group of industrial wastes that is sometimes ignored emanates from agricultural activities. These wastes fall into two categories, insecticide residue and enriched runoff from both animal manure and commercial fertilizers used on crops. The effects that insecticide residues have on fish and water fowl have been much publicized and are well known. Manure or fertilizer runoff has exactly the same effect on the environment as does sewage, and in some places it is a major component in causing or adding to eutrophication.

One other ubiquitous industrial pollutant is heat—a by-product of many industries, but especially of the electrical power industry. Some of this heat is discharged directly into the atmosphere, while some reaches the atmosphere after having first been passed into the hydrosphere. Many calculations and predictions have been made concerning the effect this heat has on weather and climate. At present the energy utilized on earth is about 7/1,000 of one percent (0.00007) of the amount received from the sun. It appears, therefore, that at the present time the weather and climate changes brought about by man's efforts to produce energy are negligible because of the relatively small amount of energy involved. Of course, this is just considering heat on a global basis. When, as sometimes happens, small enclosed bodies of water are heated by man's activities, there may be major local changes in the environment.

2105 Other Waste Products. Other types of waste products that are discharged into the atmosphere and the hydrosphere include noise, solids, and radioactive materials. Very little study has been devoted to noise pollution, and some of the basic parameters, such as critical levels, are just starting to be investigated.

Much work is being done on solid wastes. However, it should be kept in mind that occasionally, as with many other so-called pollutants, there are benefits to be derived from solid wastes, if their dumping areas are carefully chosen. For example, junked automobiles have been used successfully to attract congregations of fish. Such fishing areas develop over a relatively short period of time, as an entire ecological community collects around the obstructions, which provide hiding places for small fish and settling spots for some benthos and other creatures.

The problem of radioactive waste is not so straightforward because there are some basic questions that have not yet been answered regarding safe levels of radioactivity. Generally speaking, the average American receives, in the form of background radiation, a dosage of about 0.1 REM* per year. This intake comes from cosmic rays, from association with other people, and from the radioactivity of the earth itself. In addition, the average person receives about the same amount (0.1 REM per year) from the average of the x rays he takes in the course of his lifetime. The Atomic Energy Commission has ruled that any additional man-made radioactivity must be kept to a dosage below 0.1 REM per year for the population as a whole. At the present time, all atomic power plants are keeping their radiation levels within that limit. However, some authorities consider even this level too high and desire further research to determine what level is safe.

Nuclear power plants, of course, have a tremendous advantage in that they do not pollute the atmosphere at all in the conventional sense. The only effluents from a nuclear power plant are a slight amount of radioactivity and some heat. There are scientists who believe that, in terms of pollution from sulphur and nitrogen compounds, hydrocarbons, carbon dioxide, and carbon monoxide, the use of nuclear power is the only way man can now clean up the atmosphere. This is an interesting approach and one that must be further investigated.

2106 Managing the Environment. From the foregoing very brief summary of pollution problems it may be seen that the problems are complex. As a matter of fact, even the definition of the word *pollution* is debatable. It has been defined as any change that hurts an individual: however, one man's pollution can be another man's enrichment. It has also been defined as any measurable change in the environment: but as soon as man enters the scene there is always change.

It appears then that the problem is not so much to prevent change, but to limit it to a level that will not destroy man or his environment. It is to be hoped that, with a knowledge of the environment and a knowledge of the natural methods and processes that change it and give it its characteristics, some intelligent decision can be made as to the trade-

* Roentgen Equivalent in Man, i.e., the amount of radiation required to produce the same biological effect as one roentgen of high-penetration x rays.

offs required. And trade-offs there must be if the environment is to be properly managed, because it is impossible to return to the pristine surroundings of 5,000 years ago. All that can be expected is an environment so controlled that maximum benefits can be achieved for the maximum number of people.

Index

The Authors

Associate Professor **Jerome Williams** is Research Professor of Environmental Protection and associate chairman of the Environmental Sciences Department at the U.S. Naval Academy. He has been teaching oceanography at the Academy since the course was introduced by him in 1960.

He was graduated from the University of Maryland in 1950 with a degree in physics and received his master's degree from The Johns Hopkins University in 1952. He then remained at The Hopkins for four years doing research in underwater transparency. When he first went to the Naval Academy, in 1957, he taught electronics and physics.

Professor Williams is the author of *Oceanography*, published in 1962, *Optical Properties of the Sea*, published in 1970, *Oceanography, A First Book*, published in 1972, and *Oceanographic Instrumentation*, published in 1973. He has also published numerous papers on underwater transparency and oceanographic instrumentation. He is currently a consultant with Beckman Instruments, Inc., vice president of the Estuarine Research Federation, and a member of seven professional societies. He is a past president of the Atlantic Estuarine Research Society and is listed in *Who's Who* and *American Men and Women of Science*.

Commander **John J. Higginson,** U.S. Navy, received a bachelor of arts degree in speech and journalism from St. Mary's University in San Antonio, Texas, where he was a member of the faculty in 1955. From 1956 to 1964 he served aboard the attack transport USS *Renville* (APA-227), Tactical Air Control Squadron Thirteen, and Helicopter ASW Squadron Two. Commander Higginson earned his bachelor of science degree in meteorology with graduate work in oceanography from the U.S. Naval Postgraduate School, Monterey, California, in 1966. He was a member of the faculty at the U.S. Naval Academy where he was director of the air-ocean environment course until 1968. In 1968, he was awarded a master of science degree in management by George Washington University. From 1968 to 1972 he was operations officer for Helicopter ASW Squadron Four and the helicopter carrier USS *Okinawa* (LPH-3). During this period he participated in the recovery operations for the lunar missions Apollo Eight, Eleven, Twelve, Thirteen, and Fifteen. Commander Higginson is presently executive officer of Helicopter ASW Squadron Two at NAS Imperial Beach, California.

Commander **John D. Rohrbough,** U.S. Navy, a 1958 graduate of the U.S. Naval Academy, has spent eight years at sea as a surface warfare officer. With a master's degree in air-ocean environment from the U.S. Naval Postgraduate School at Monterey, California, Commander Rohrbough has served at the Naval Academy as instructor of oceanography and on the staff of the Superintendent. Commander Rohrbough is a graduate of the Armed Forces Staff College and is presently attached to the Organization of the Joint Chiefs of Staff.

The text of this book is set in eleven point Times Roman Monophoto with two points of leading.

The book is printed offset on 60-pound Banta Pigmented Offset. The book is bound in Holliston Blueback White.

Designed by David Q. Scott.
Line drawings by Bill Clipson

The book was composed, printed, and bound by the George Banta Company, Menasha, Wisconsin.

